PROBLEMS & SOLUTIONS IN THEORETICAL & MATHEMATICAL PHYSICS

PROBLEMS & SOLUTIONS IN THEORETICAL & MATHEMATICAL PHYSICS

Second Edition

Volume I: Introductory Level

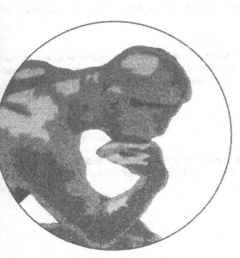

Willi-Hans Steeb

Rand Afrikaans University, South Africa

World Scientific

New Jersey • London • Singapore • Hong Kong

Published by

World Scientific Publishing Co. Pte. Ltd.

5 Toh Tuck Link, Singapore 596224

USA office: Suite 202, 1060 Main Street, River Edge, NJ 07661

UK office: 57 Shelton Street, Covent Garden, London WC2H 9HE

British Library Cataloguing-in-Publication Data
A catalogue record for this book is available from the British Library.

PROBLEMS AND SOLUTIONS IN THEORETICAL AND MATHEMATICAL PHYSICS
Vol. 1: Introductory Level (Second Edition)

ISBN 981-238-990-3
ISBN 981-238-989-X (pbk)

This book is printed on acid-free paper.

Printed in Singapore by Uto-Print

Preface

The purpose of this book is to supply a collection of problems together with their detailed solution which will prove to be valuable to students as well as to research workers in the fields of mathematics, physics, engineering and other sciences. The topics range in difficulty from elementary to advanced. Almost all problems are solved in detail and most of the problems are self-contained. All relevant definitions are given. Students can learn important principles and strategies required for problem solving. Teachers will also find this text useful as a supplement, since important concepts and techniques are developed in the problems. The material was tested in my lectures given around the world.

The book is divided into two volumes. Volume I presents the introductory problems for undergraduate and advanced undergraduate students. In Volume II the more advanced problems together with their detailed solutions are collected, to meet the needs of graduate students and researchers. Problems included cover most of the new fields in theoretical and mathematical physics such as Lax representation, Bäcklund transformation, soliton equations, Lie algebra valued differential forms, Hirota technique, Painlevé test, the Bethe ansatz, the Yang-Baxter relation, wavelets, chaos, fractals, complexity, etc.

In the reference section some other books are listed having useful problems for students in theoretical physics and mathematical physics. Some or related problems to those given in volume I can be found in these books. In Volume II references are given to books and original articles where some of the advanced problems can be found.

I wish to express my gratitude to Catharine Thompson and Lance Hoffman for a critical reading of the manuscripts. Finally, I appreciate the help of the Lady from Madeira.

Any useful suggestions and comments are welcome.

Email addresses of the author:

```
steeb_wh@yahoo.com
whs@na.rau.ac.za
Willi-Hans.Steeb@fhso.ch
```

Home pages of the author:

```
http://issc.rau.ac.za
http://zeus.rau.ac.za/steeb/steeb.html
http://www.fhso.ch/~steeb
```

Contents

vii

Notation

	empty set
N	natural numbers
	integers
	rational numbers
	real numbers
$+$	nonnegative real numbers
	complex numbers
^{n}	n-dimensional Euclidian space
^{n}	n-dimensional complex linear space
	$:= \sqrt{-1}$
z	real part of the complex number z
z	imaginary part of the complex number z
$\in \mathbf{R}^n$	element $\mathbf{x}$ of $\mathbf{R}^n$
$\subset B$	subset A of set B
$\cap B$	the intersection of the sets A and B
$\cup B$	the union of the sets A and B
$\circ g$	composition of two mappings $(f \circ g)(x) = f(g(x))$
	dependent variable
	independent variable (time variable)
	independent variable (space variable)
$^T = (x_1, x_2, \ldots, x_n)$	vector of independent variables, ^{T} means transpose
$^T = (u_1, u_2, \ldots, u_n)$	vector of dependent variables, ^{T} means transpose
$\cdot\|$	norm
$\cdot \mathbf{y}$	scalar product (inner product)

$\mathbf{x} \times \mathbf{y}$	vector product
$\otimes$	Kronecker product, tensor product
det	determinant of a square matrix
tr	trace of a square matrix
I	unit matrix
$[\,,\,]$	commutator
δ_{jk}	Kronecker delta with $\delta_{jk} = 1$ for $j = k$ and $\delta_{jk} = 0$ for $j \neq k$
$\mathrm{sgn}(x)$	the sign of x, 1 if $x > 0$, -1 if $x < 0$, 0 if $x = 0$
λ	eigenvalue
ϵ	real parameter
$\wedge$	Grassmann product (exterior product, wedge product)
H	Hamilton function
$\hat{H}$	Hamilton operator

Chapter 1

Complex Numbers and Functions

Problem 1. (i) Let $i := \sqrt{-1}$. Calculate

$$i^i.$$

(ii) Let $z = x + iy$ with $x, y \in \mathbf{R}$. Calculate

$$|e^{iz}|.$$

Solution. (i) Let z_1 and z_2 be complex numbers. Assume that $z_1 \neq 0$. One defines

$$z_1^{z_2} := e^{z_2 \ln z_1}.$$

Now

$$\ln z = \ln r + i(\theta + 2k\pi) \quad k \in \mathbf{Z}$$

where $z = x + iy$, $x, y \in \mathbf{R}$ and

$$r := \sqrt{x^2 + y^2}.$$

Therefore $\ln z$ is an infinitely many valued function. The *principal branch* of $\ln z$ is defined as $\ln r + i\theta$ where $0 \leq \theta < 2\pi$. Consequently, for the principal branch we have

$$i^i = e^{i \ln i} = e^{ii\pi/2} = e^{-\pi/2}.$$

(ii) Since $z = x + iy$ with $x, y \in \mathbf{R}$, the complex conjugate is given by $\bar{z} = x - iy$. Therefore

$$|e^{iz}| := \sqrt{e^{iz} e^{-i\bar{z}}} = \sqrt{e^{i(x+iy)} e^{-i(x-iy)}} = \sqrt{e^{-2y}} = e^{-y}.$$

Problem 2.   Let

$$z = x + iy$$

where $x, y \in \mathbf{R}$. Find the real and imaginary part of

$$\cos\left(\frac{z}{2}\right) \sin\left(\frac{z}{2}\right).$$

Solution.   Since

$$\cos\left(\frac{z}{2}\right) \equiv \frac{e^{i(x+iy)/2} + e^{-i(x+iy)/2}}{2}$$

and

$$\overline{\sin\left(\frac{z}{2}\right)} \equiv \sin\left(\frac{\bar{z}}{2}\right) \equiv \frac{e^{i(x-iy)/2} - e^{-i(x-iy)/2}}{2i},$$

we obtain

$$\cos\left(\frac{z}{2}\right) \sin\left(\frac{\bar{z}}{2}\right) \equiv \frac{e^{ix} - e^{-ix} + e^{y} - e^{-y}}{4i}$$

or

$$\cos\left(\frac{z}{2}\right) \sin\left(\frac{\bar{z}}{2}\right) \equiv \frac{1}{2}\sin x + \frac{1}{2i}\sinh y \equiv \frac{1}{2}\sin x - \frac{i}{2}\sinh y.$$

Consequently, the real part is

$$\frac{1}{2}\sin x$$

and the imaginary part is

$$-\frac{1}{2}\sinh y.$$

Problem 3.   Let n be an integer. Then

$$e^{1+2n\pi i} = e.$$

If we write

$$(e^{1+2n\pi i})^{1+2n\pi i} = e^{1+2n\pi i} = e \qquad (1)$$

and

$$(e^{1+2n\pi i})^{1+2n\pi i} = e^{1+4n\pi i - 4n^2\pi^2} = ee^{-4n^2\pi^2}, \qquad (2)$$

t follows that

$$e^{-4n^2\pi^2} = 1.$$

Solve this paradox.

Solution. Define the *imaginary remainder* $\Im r(z)$ and the *imaginary quotient* $\Im q(z)$ by

$$\Im(z) = \Im r(z) + 2\pi \Im q(z),$$

where

$$\Im r(z) \in (-\pi, \pi]$$

nd

$$\Im q(z) \in \mathbf{Z}.$$

We have

$$(e^u)^v = e^{uv} e^{-v2\pi i \Im q(u)}$$

nce

$$(e^u)^v = e^{\log(e^u)^v} = e^{(\Re(u) + i\Im r(u))v} = e^{(u - i2\pi \Im q(u))v} = e^{uv} e^{-vi2\pi \Im q(u)}.$$

1 (2), we said that

$$(e^{1+2n\pi i})^{1+2n\pi i} = e^{1+4n\pi i - 4n^2\pi^2}.$$

his should be replaced by

$$(e^{1+2n\pi i})^{1+2n\pi i} = e^{1+4n\pi i - 4n^2\pi^2} e^{-(1+2n\pi i)2\pi i \Im q(1+2n\pi i)}$$

$$= e^{1-4n^2\pi^2} e^{-(1+2n\pi i)2\pi i n} = e^{1-4n^2\pi^2} e^{-2n\pi i + 4n^2\pi^2} = e$$

hich agrees with (1).

emark   *In Problem 1, we considered the domain* $[0, 2\pi)$ *for the principal* ranch. In this problem we consider the domain $(-\pi, \pi]$ for the principal ranch.

Problem 4.   Show that

$$\cos(2\theta) \equiv \cos^2 \theta - \sin^2 \theta\,, \tag{1}$$

$$\sin(2\theta) \equiv 2 \sin \theta \cos \theta \tag{2}$$

using De Moivre's Formula, where $\theta \in \mathbf{R}$.

Solution.   Let n be a positive integer. Then

$$z^n = (re^{i\theta})^n = r^n e^{in\theta} = r^n(\cos(n\theta) + i\sin(n\theta))\,.$$

If

$$|z| = r = 1\,,$$

then we have

$$e^{in\theta} \equiv \cos(n\theta) + i\sin(n\theta) \equiv (\cos\theta + i\sin\theta)^n\,.$$

This is know as *De Moivre's theorem*. Now let $n = 2$. Then it follows that

$$\cos(2\theta) + i\sin(2\theta) = e^{2i\theta} = (e^{i\theta})^2 = (\cos\theta + i\sin\theta)^2$$

$$= \cos^2\theta - \sin^2\theta + 2i\sin\theta\cos\theta\,.$$

The real and imaginary parts of which give the trigonometric relations (1) and (2).

Remark.   *We can find trigonometric identities for arbitrary n. For example for $n = 3$, we find the identities*

$$\cos(3\theta) \equiv 4\cos^3\theta - 3\cos\theta\,,$$

$$\sin(3\theta) \equiv 3\sin\theta - 4\sin^3\theta$$

by considering the real and imaginary parts.

Chapter 2

Sums and Product

Problem 1. (i) Calculate

$$S = \sum_{n=1}^{\infty} \frac{1}{3^n} \, . \tag{1}$$

(ii) Calculate

$$S(n, x) = \sum_{k=0}^{n} x^k$$

where $x \neq 0$. The two series are so-called *geometric series*.

Solution. (i) We can use the ratio test or the nth root test to prove that the series (1) converges absolutely. We have

$$S = \sum_{n=1}^{\infty} \frac{1}{3^n} \equiv \frac{1}{3} + \frac{1}{9} + \frac{1}{27} + \frac{1}{81} + \cdots .$$

$$\equiv \frac{1}{3} \left(1 + \frac{1}{3} + \frac{1}{9} + \frac{1}{27} + \cdots \right) \equiv \frac{1}{3}(1 + S).$$

Consequently,

$$S = \frac{1}{2} \, .$$

(ii) Let $x = 1$. Then obviously

$$S(n, 1) = \sum_{k=0}^{n} 1 = n + 1 \, .$$

5

Now let $x \neq 1$. Since

$$S(n, x) = \sum_{k=0}^{n} x^k = 1 + x + x^2 + \cdots + x^n, \tag{2}$$

we obtain

$$xS(n, x) = \sum_{k=0}^{n} x^{k+1} = x + x^2 + \cdots + x^n + x^{n+1}. \tag{3}$$

Subtracting (2) from (3) gives

$$(x - 1)S(n, x) = -1 + x^{n+1}$$

or

$$S(n, x) = \frac{1 - x^{n+1}}{1 - x}.$$

Problem 2.   Let $n \in \mathbf{N}$. Evaluate the sum

$$\sum_{k=1}^{n} \frac{k^2}{2^k}. \tag{1}$$

Solution.   If we define

$$S(x) := \sum_{k=1}^{n} k^2 x^k,$$

then the sum (1) is given by $S(\frac{1}{2})$. Now we can use the techniques of analysis. We know that

$$\sum_{k=0}^{n} x^k = \frac{1 - x^{n+1}}{1 - x}, \quad x \neq 1.$$

Differentiating each side with respect to x yields

$$\sum_{k=1}^{n} kx^{k-1} = \frac{1 - (n+1)x^n + nx^{n+1}}{(1 - x)^2}.$$

Multiplying each side of this equation by x, differentiating a second time, and multiplying the result by x yields

$$S(x) = \sum_{k=1}^{n} k^2 x^k = \frac{x(1+x) - x^{n+1}(nx - n - 1)^2 - x^{n+2}}{(1 - x)^3}. \tag{2}$$

From (2), it follows that

$$S\left(\frac{1}{2}\right) = \sum_{k=1}^{n} \frac{k^2}{2^k} = 6 - \frac{1}{2^{n-2}}\left(\frac{1}{2}n - n - 1\right)^2 - \frac{1}{2^{n-1}} = 6 - \left(\frac{n^2 + 4n + 6}{2^n}\right).$$

$$(3)$$

Problem 3.   Calculate

$$S(\alpha) = \sum_{n=0}^{N-1} e^{in\alpha}, \qquad (1)$$

where $\alpha \in \mathbf{R}$ and $N \in \mathbf{N}$.

Solution.   Let $\alpha = 2m\pi$ where $m \in \mathbf{Z}$. Since

$$e^{2inm\pi} = 1, \quad n \in \mathbf{N} \cup \{0\}$$

we find

$$S(2m\pi) = N.$$

Now let $\alpha \neq 2m\pi$, where $m \in \mathbf{Z}$. Owning to $e^{in\alpha} \equiv (e^{i\alpha})^n$ the sum (1) over n is a geometric series. Consequently,

$$S(\alpha) = \sum_{n=0}^{N-1} e^{in\alpha} = \frac{1 - e^{iN\alpha}}{1 - e^{i\alpha}} = \frac{\sin(N\alpha/2)}{\sin(\alpha/2)} e^{i(N-1)\alpha/2}.$$

Remark.   *The sums*

$$\sum_{n=1}^{N-1} n e^{in\alpha}$$

and

$$\sum_{n=1}^{N-1} n^2 e^{in\alpha}$$

can now easily be calculated since

$$\frac{dS}{d\alpha} = i \sum_{n=1}^{N-1} n e^{in\alpha}$$

and

$$\frac{d^2 S}{d\alpha^2} = -\sum_{n=1}^{N-1} n^2 e^{in\alpha}.$$

Problem 4.   Calculate

$$\sum_{k=-N}^{N} \frac{1}{1+e^{\lambda\epsilon(k)}} \tag{1}$$

where $N \in \mathbf{N}$,

$$\epsilon(-k) = -\epsilon(k)$$

and λ is a real parameter.

Solution.   We define

$$S(N,\lambda) := \sum_{k=-N}^{N} \frac{1}{1+e^{\lambda\epsilon(k)}}\,. \tag{2}$$

Taking the derivative of S with respect to λ, we obtain

$$\frac{dS}{d\lambda} = -\sum_{k=-N}^{N} \frac{\epsilon(k)e^{\lambda\epsilon(k)}}{(1+e^{\lambda\epsilon(k)})^2} \equiv -\sum_{k=-N}^{N} \frac{\epsilon(k)}{2+2\cosh(\lambda\epsilon(k))}\,.$$

Since $\epsilon(-k) = -\epsilon(k)$ and $\cosh(x)$ is an even function we find

$$\frac{dS}{d\lambda} = 0\,. \tag{3}$$

From (2), we have

$$S(N,\lambda=0) = \sum_{k=-N}^{N} \frac{1}{1+1} = \frac{1}{2}(2N+1)\,. \tag{4}$$

Consequently, the solution of the linear differential Eq. (3) with the initial condition (4) is given by

$$S(N,\lambda) = N + \frac{1}{2}\,.$$

This means the sum (1) is independent of λ.

Problem 5.   Let

$$z = \sum_{n\in I} a_n e^{-ix_n} \tag{1}$$

where I is a finite set and $a_n,\ x_n \in \mathbf{R}$. Calculate $z\bar{z}$, where $\bar{z}$ denotes the complex conjugate of z.

Solution. From (1), we obtain

$$\bar{z} = \sum_{m \in I} a_m e^{ix_m} .$$

Therefore

$$z\bar{z} = \sum_{n \in I} \sum_{m \in I} a_n a_m e^{-ix_n} e^{ix_m} = \sum_{n \in I} \sum_{m \in I} a_n a_m e^{i(x_m - x_n)} .$$

Thus

$$z\bar{z} = \sum_{n \in I} \sum_{m \in I} a_n a_m (\cos(x_m - x_n) + i \sin(x_m - x_n)) .$$

Since

$$\sin(-\alpha) \equiv -\sin(\alpha) ,$$

we have

$$\sin(x_m - x_n) \equiv \sin(-(x_n - x_m)) \equiv -\sin(x_n - x_m) .$$

Therefore

$$\sum_{n \in I} \sum_{m \in I} a_n a_m \sin(x_m - x_n) = 0 .$$

Consequently,

$$z\bar{z} = \sum_{n \in I} \sum_{m \in I} a_n a_m \cos(x_m - x_n) .$$

Problem 6. Let a be a positive constant. Let

$$S := \left\{ \frac{a}{2}(1,1,1), \frac{a}{2}(1,1,-1), \frac{a}{2}(1,-1,1), \frac{a}{2}(-1,1,1), \right.$$

$$\left. \frac{a}{2}(1,-1,-1), \frac{a}{2}(-1,1,-1), \frac{a}{2}(-1,-1,1), \frac{a}{2}(-1,-1,-1) \right\} .$$

Calculate

$$\sum_{\Delta \in S} e^{i(\mathbf{k} \cdot \boldsymbol{\Delta})} ,$$

where $\mathbf{k} \cdot \boldsymbol{\Delta} := k_1 \Delta_1 + k_2 \Delta_2 + k_3 \Delta_3$.

Solution.

$$\sum_{\Delta \in S} e^{i(\mathbf{k} \cdot \Delta)} \equiv \sum_{\Delta \in S} e^{i(k_1 \Delta_1 + k_2 \Delta_2 + k_3 \Delta_3)}$$

$$\equiv \left(e^{i(k_1 a/2 + k_2 a/2 + k_3 a/2)} + e^{-i(k_1 a/2 + k_2 a/2 + k_3 a/2)} \right.$$

$$+ e^{i(k_1 a/2 + k_2 a/2 - k_3 a/2)} + e^{-i(k_1 a/2 + k_2 a/2 - k_3 a/2)}$$

$$+ e^{i(k_1 a/2 - k_2 a/2 + k_3 a/2)} + e^{-i(k_1 a/2 - k_2 a/2 + k_3 a/2)}$$

$$\left. + e^{i(-k_1 a/2 + k_2 a/2 + k_3 a/2)} + e^{-i(-k_1 a/2 + k_2 a/2 + k_3 a/2)} \right).$$

Thus

$$\sum_{\Delta \in S} e^{i\mathbf{k} \cdot \Delta} \equiv 2 \left[\cos \left(\frac{k_1 a}{2} + \frac{k_2 a}{2} + \frac{k_3 a}{2} \right) + \cos \left(\frac{k_1 a}{2} + \frac{k_2 a}{2} - \frac{k_3 a}{2} \right) \right.$$

$$\left. + \cos \left(\frac{k_1 a}{2} - \frac{k_2 a}{2} + \frac{k_3 a}{2} \right) + \cos \left(-\frac{k_1 a}{2} + \frac{k_2 a}{2} + \frac{k_3 a}{2} \right) \right].$$

Since for b, c, $d \in \mathbf{R}$, we have the identity

$$\cos(b + c + d) + \cos(b + c - d) + \cos(b - c + d) + \cos(-b + c + d)$$

$$\equiv 4 \cos(b) \cos(c) \cos(d),$$

we obtain

$$\sum_{\Delta \in S} e^{i(\mathbf{k} \cdot \Delta)} \equiv 8 \cos \left(\frac{k_1 a}{2} \right) \cos \left(\frac{k_2 a}{2} \right) \cos \left(\frac{k_3 a}{2} \right).$$

Problem 7.   (i) Let

$$A_j := \frac{1}{x - a_j}, \qquad B_{kj} := \frac{1}{a_k - a_j}$$

where $k \neq j$. Show that

$$A_n B_{ns} + A_s B_{sn} = A_n A_s. \tag{1}$$

(ii) Show that the identity

$$\sum_{\substack{j,k=1 \\ j \neq k}}^{n} \frac{1}{x - a_j} \frac{1}{x - a_k} \equiv 2 \sum_{\substack{j,k=1 \\ j \neq k}}^{n} \frac{1}{x - a_k} \frac{1}{a_k - a_j} \tag{2}$$

holds. Using the identity (1), we can write (2) as

$$\sum_{\substack{j,k=1 \\ j \neq k}}^{n} A_j A_k \equiv 2 \sum_{\substack{j,k=1 \\ j \neq k}}^{n} A_k B_{kj}.$$

Solution. (i) By straightfoward calculation we find

$$A_n B_{ns} + A_s B_{sn} = \frac{1}{x - a_n} \frac{1}{a_n - a_s} + \frac{1}{x - a_s} \frac{1}{a_s - a_n}$$

$$= \frac{1}{a_n - a_s} \left(\frac{1}{x - a_n} - \frac{1}{x - a_s} \right)$$

$$= \frac{1}{a_n - a_s} \left(\frac{x - a_s - x + a_n}{(x - a_n)(x - a_s)} \right)$$

$$= \frac{1}{(x - a_n)(x - a_s)}$$

$$= A_n A_s.$$

(ii) We find that

$$\sum_{\substack{j,k=1 \\ j \neq k}}^{n} A_j A_k \equiv \sum_{\substack{j,k=1 \\ j \neq k}}^{n-1} A_j A_k + \sum_{s=1}^{n-1} A_n A_s + \sum_{t=1}^{n-1} A_t A_n.$$

Using (1), we obtain

$$\sum_{\substack{j,k=1 \\ j \neq k}}^{n} A_j A_k \equiv 2 \sum_{\substack{j,k=1 \\ j \neq k}}^{n-1} A_k B_{kj} + 2 A_n \sum_{s=1}^{n-1} A_s$$

and

$$A_n \sum_{s=1}^{n-1} A_s \equiv \sum_{s=1}^{n-1} (A_n B_{ns} + A_s B_{sn}) \equiv \sum_{s=1}^{n-1} A_n B_{ns} + \sum_{t=1}^{n-1} A_t B_{tn}.$$

Therefore identity (2) follows.

Problem 8. Let ϵ be a real parameter and $a_n, b_n \in \mathbf{R}$. Assume that

$$\exp \left(\sum_{n=1}^{\infty} \frac{(i\epsilon)^n b_n}{n!} \right) = \sum_{n=0}^{\infty} \frac{(i\epsilon)^n a_n}{n!} \tag{1}$$

where $a_0 = 1$. Find the relationship between the coefficients a_n and b_n.

Solution.   An arbitrary term of the exponential function on the left hand side of (1) is given by

$$\frac{1}{k!}\left(\sum_{n=1}^{\infty}\frac{(i\epsilon)^n b_n}{n!}\right)^k = \frac{1}{k!}\left(\sum_{n_1=1}^{\infty}\frac{(i\epsilon)^{n_1} b_{n_1}}{n_1!}\right)\cdots\left(\sum_{n_k=1}^{\infty}\frac{(i\epsilon)^{n_k} b_{n_k}}{n_k!}\right)$$

$$= \frac{1}{k!}\sum_{n_1=1}^{\infty}\sum_{n_2=1}^{\infty}\cdots\sum_{n_k=1}^{\infty}\frac{(i\epsilon)^{n_1+n_2+\cdots+n_k} b_{n_1} b_{n_2}\cdots b_{n_k}}{n_1! n_2!\cdots n_k!}.$$

Therefore

$$\exp\left(\sum_{n=1}^{\infty}\frac{(i\epsilon)^n b_n}{n!}\right)$$

$$\equiv 1 + \sum_{n=1}^{\infty}\frac{(i\epsilon)^n b_n}{n!} + \frac{1}{2!}\sum_{n_1=1}^{\infty}\sum_{n_2=1}^{\infty}\frac{(i\epsilon)^{n_1+n_2} b_{n_1} b_{n_2}}{n_1! n_2!} + \cdots$$

$$+ \frac{1}{k!}\sum_{n_1=1}^{\infty}\sum_{n_2=1}^{\infty}\cdots\sum_{n_k=1}^{\infty}\frac{(i\epsilon)^{n_1+n_2+\cdots+n_k} b_{n_1} b_{n_2}\cdots b_{n_k}}{n_1! n_2!\cdots n_k!} + \cdots$$

$$= \sum_{n=0}^{\infty}\frac{(i\epsilon)^n a_n}{n!}.$$

Equating terms of the same power in $i\epsilon$, we obtain for the first three terms

$$(i\epsilon)^1 : \quad a_1 = b_1,$$
$$(i\epsilon)^2 : \quad a_2 = b_2 + b_1^2,$$
$$(i\epsilon)^3 : \quad a_3 = b_3 + 3b_2 b_1 + b_1^3.$$

It follows that

$$b_1 = a_1, \quad b_2 = a_2 - a_1^2, \quad b_3 = a_3 - 3a_2 a_1 + 2a_1^3.$$

Remark.   *Let X be a random variable with probability density function $f_X(x)$. The n-th moment of X is defined as*

$$\langle X^n\rangle := \int x^n f_X(x)dx$$

where the integral is over the entire range of X and $n = 0, 1, 2, \ldots$ Therefore

$$\langle X^0\rangle = 1.$$

The characteristic function $\phi_X(k)$ *is defined as*

$$\phi_X(k) := \sum_{n=0}^{\infty} \frac{(ik)^n \langle X^n \rangle}{n!}. \tag{2}$$

The cumulant expansion *is given by*

$$\phi_X(k) = \exp\left(\sum_{n=1}^{\infty} \frac{(ik)^n C_n(X)}{n!}\right). \tag{3}$$

Thus the above Eqs. (2) and (2) give the connection between $C_n(X)$ and $\langle X^n \rangle$. The cumulant expansion also plays an important role in high temperature expansion in statistical physics.

Problem 9.   Show that

$$(\cos\theta + i\sin\theta)^n \equiv \cos(n\theta) + i\sin(n\theta) \tag{1}$$

where $\theta \in \mathbf{R}$ and $n \in \mathbf{N}$.

Remark.   *Identity (1) is sometimes called* De Moivre's theorem.

Solution .   We apply the *principle of mathematical induction*.   The identity (1) is clearly true for $n = 1$.   Assume that the identity is true for k, i.e.,

$$(\cos\theta + i\sin\theta)^k \equiv \cos(k\theta) + i\sin(k\theta).$$

Multiplying both sides by $\cos\theta + i\sin\theta$ yields

$$(\cos\theta + i\sin\theta)^{k+1} = (\cos k\theta + i\sin k\theta)(\cos\theta + i\sin\theta)$$

or

$$(\cos\theta + i\sin\theta)^{k+1} = \cos(k+1)\theta + i\sin(k+1)\theta.$$

Consequently, if the result is true for

$$n = k,$$

then it is also true for

$$n = k+1.$$

Since the result is true for $n = 1$, it has to be true for $n = 1 + 1 = 2$, $n = 2 + 1$ etc. and so has to be true for all $n \in \mathbf{N}$.

Remark.   *We have*

$$(e^{i\theta})^n \equiv e^{in\theta}$$

where $\theta \in \mathbf{R}$ and $n \in \mathbf{N}$.

Problem 10.   Let f_k ($k = 1, 2, \ldots, n$) be differentiable functions. Assume that $f_k \neq 0$ for $k = 1, 2, \ldots, n$. Let

$$S(x) = \prod_{k=1}^{n} f_k(x) \equiv f_1(x) f_2(x) \cdots f_n(x).$$

Calculate the derivative of S with respect to x.

Solution.   We apply the *product rule*

$$\frac{d}{dx}(h(x)g(x)) = \frac{dh}{dx}g + h\frac{dg}{dx}$$

to

$$\frac{dS}{dx} = \frac{d}{dx}\prod_{k=1}^{n} f_k(x).$$

Thus

$$\frac{dS}{dx} \equiv \frac{df_1}{dx}f_2\cdots f_n + f_1\frac{df_2}{dx}\cdots f_n + \cdots + f_1 f_2 \cdots \frac{df_n}{dx}$$

or

$$\frac{dS}{dx} \equiv \frac{1}{f_1}\frac{df_1}{dx}S(x) + \frac{1}{f_2}\frac{df_2}{dx}S(x) + \cdots + \frac{1}{f_n}\frac{df_n}{dx}S(x).$$

Consequently,

$$\frac{dS}{dx} = S(x)\sum_{j=1}^{n}\frac{1}{f_j}\frac{df_j}{dx} \equiv S(x)\sum_{j=1}^{n}\frac{d}{dx}(\ln f_j(x)).$$

Problem 11.   Show that every real number $r > 0$ can be represented as the *Cantor series*

$$r = \sum_{\nu=1}^{\infty}\frac{c_\nu}{\nu!} = c_1 + \frac{c_2}{2!} + \frac{c_3}{3!} + \cdots \tag{1}$$

where

$$0 \leq c_\nu \leq \nu - 1, \quad \nu > 1, \quad c_\nu \in \mathbf{N} \cup \{0\}.$$

Solution. We write the real number r in the form

$$r = [r] + \frac{\rho_2}{2} = c_1 + \frac{\rho_2}{2}, \tag{2}$$

where $[r]$ denotes the largest integer $\leq r$. Thus $[r] = c_1$ and $\rho_2 < 2$. We set

$$\rho_2 = [\rho_2] + \frac{\rho_3}{3} = c_2 + \frac{\rho_3}{3}, \quad \rho_{n-1} = [\rho_{n-1}] + \frac{\rho_n}{n} = c_{n-1} + \frac{\rho_n}{n} \tag{3}$$

and

$$\rho_n = [\rho_n] + \frac{\rho_{n+1}}{n+1} = c_n + \frac{\rho_{n+1}}{n+1}. \tag{4}$$

Thus we have

$$c_m \leq \rho_m < m \tag{5}$$

for $m = 2, 3, 4, \ldots, n$. Inserting (3) and (4) into (2) yields

$$r = c_1 + \frac{c_2}{2} + \frac{c_3}{2 \cdot 3} + \ldots + \frac{c_n}{n!} + \frac{\rho_{n+1}}{(n+1)!}.$$

Using (5), we find

$$0 \leq r - \left(c_1 + \frac{c_2}{2} + \frac{c_3}{3!} + \cdots + \frac{c_n}{n!} \right) < \frac{1}{n!}.$$

Thus (1) follows.

Chapter 3

Discrete Fourier Transform

Problem 1. (i) Calculate

$$\sum_{k=0}^{N-1} e^{i2\pi k(n-m)/N} \tag{1}$$

where $n, m \in \mathbf{N} \cup \{0\}$, $N \in \mathbf{N}$ and $0 \le n, m \le N - 1$.

(ii) Let

$$\hat{x}(k) := \frac{1}{N} \sum_{n=0}^{N-1} x(n)e^{-i2\pi kn/N} \tag{2}$$

where $N \in \mathbf{N}$ and $k = 0, 1, 2, \ldots, N - 1$. From (2), find $x(n)$, i.e., find the discrete inverse Fourier transform.

Remark. $\hat{x}(k)$ *is called the* discrete Fourier transform *in one dimension and* $x(0)$, $x(1), \ldots, x(N - 1)$ *are N data points.*

Solution. (i) If $n = m$, then

$$\sum_{k=0}^{N-1} 1 = N.$$

If $n \ne m$, then $n - m = q$, where $0 < q \le N - 1$. Then the sum (1) is a geometric series.

Therefore

$$\sum_{k=0}^{N-1} e^{i2\pi kq/N} = 0.$$

Consequently,

$$\sum_{k=0}^{N-1} e^{i2\pi k(n-m)/N} = N\delta_{nm} \tag{3}$$

where δ_{nm} denotes the *Kronecker delta*.

(ii) From (2), it follows that

$$\hat{x}(k)e^{i2\pi km/N} = \frac{1}{N}\sum_{n=0}^{N-1} x(n)e^{i2\pi k(m-n)/N} .$$

Summation over k of both sides gives

$$\sum_{k=0}^{N-1}\hat{x}(k)e^{i2\pi km/N} = \frac{1}{N}\sum_{k=0}^{N-1}\sum_{n=0}^{N-1} x(n)e^{i2\pi k(m-n)/N}$$

$$= \frac{1}{N}\sum_{n=0}^{N-1} x(n)\sum_{k=0}^{N-1} e^{i2\pi k(m-n)/N} .$$

Using (3) we obtain

$$\sum_{k=0}^{N-1}\hat{x}(k)e^{i2\pi km/N} = \frac{1}{N}\sum_{n=0}^{N-1} x(n)N\delta_{mn} = x(m) .$$

Therefore

$$x(n) = \sum_{k=0}^{N-1}\hat{x}(k)e^{i2\pi kn/N} .$$

Problem 2.   Let

$$x(n) = \cos\left(\frac{2\pi n}{N}\right)$$

where $N = 8$ and $n = 0, 1, 2, \ldots, N - 1$. Find $\hat{x}(k)$ $(k = 0, 1, 2, \ldots, N - 1)$, i.e., find the discrete Fourier transform.

Solution.   We have

$$\hat{x}(k) = \frac{1}{8}\sum_{n=0}^{7}\cos\left(\frac{2\pi n}{8}\right)e^{-i2\pi kn/8} .$$

Using the identity

$$\cos\left(\frac{2\pi n}{8}\right) \equiv \frac{e^{i2\pi n/8} + e^{-i2\pi n/8}}{2}$$

we have

$$\hat{x}(k) = \frac{1}{16}\sum_{n=0}^{7}(e^{i2\pi n(1-k)/8} + e^{-i2\pi n(1+k)/8}).$$

Consequently,

$$\hat{x}(k) = \begin{cases} \dfrac{1}{2} & \text{for } k = 1 \\ \dfrac{1}{2} & \text{for } k = 7 \\ 0 & \text{otherwise} \end{cases}.$$

Problem 3.   Let $\mathbf{x} = (x(0), x(1), \ldots, x(N-1))$ and $\hat{\mathbf{x}} = (\hat{x}(0), \hat{x}(1), \ldots, \hat{x}(N-1))$. Define

$$(\mathbf{x}, \mathbf{x}) := \sum_{n=0}^{N-1} \bar{x}(n)x(n), \quad (\hat{\mathbf{x}}, \hat{\mathbf{x}}) := \sum_{k=0}^{N-1} \bar{\hat{x}}(k)\hat{x}(k) \tag{1}$$

where $\bar{x}(n)$ denotes the complex conjugate of $x(n)$. Show that

$$(\mathbf{x}, \mathbf{x}) = N(\hat{\mathbf{x}}, \hat{\mathbf{x}}).$$

Solution.   From (1), we have

$$(\mathbf{x}, \mathbf{x}) = \sum_{n=0}^{N-1} \bar{x}(n)x(n) = \sum_{k=0}^{N-1}\sum_{l=0}^{N-1} \bar{\hat{x}}(k)\hat{x}(l) \sum_{n=0}^{N-1} e^{i2\pi n(l-k)/N}.$$

Using

$$\sum_{n=0}^{N-1} e^{i2\pi n(l-k)/N} = N\delta_{lk},$$

we find

$$(\mathbf{x}, \mathbf{x}) = N\sum_{k=0}^{N-1} \bar{\hat{x}}(k)\hat{x}(k).$$

Thus

$$(\mathbf{x}, \mathbf{x}) = N(\hat{\mathbf{x}}, \hat{\mathbf{x}}).$$

Chapter 4

Algebraic and Transcendental Equations

Problem 1. (i) Let $n \in \mathbf{N}$. Solve

$$z^n = 1.$$ (1)

(ii) Find all solutions of

$$x^4 + x^3 + x^2 + x + 1 = 0.$$ (2)

Solution. (i) Since every complex number z can be written as

$$z = r e^{i\phi},$$

we have

$$z^n = r^n e^{in\phi}.$$

Taking (1) into account, we find $r = 1$ and

$$e^{in\phi} \equiv \cos(n\phi) + i\sin(n\phi) = 1.$$

Therefore

$$z_k = \cos\left(\frac{2\pi k}{n}\right) + i\sin\left(\frac{2\pi k}{n}\right)$$

are the roots of 1, where $k = 0, 1, \ldots, n-1$.

(ii) Method 1. Equation (2) can be solved by dividing by x^2, substituting $y = x + 1/x$, and then applying the quadratic formula. Thus, we have

$$x^2 + \frac{1}{x^2} + x + \frac{1}{x} + 1 = 0,$$

19

$$\left(x^2 + 2 + \frac{1}{x^2}\right) + \left(x + \frac{1}{x}\right) + (1 - 2) = 0,$$

$$\left(x + \frac{1}{x}\right)^2 + \left(x + \frac{1}{x}\right) - 1 = 0,$$

$$y^2 + y - 1 = 0.$$

The roots of this equation are

$$y_1 = \frac{-1 + \sqrt{5}}{2}, \quad y_2 = \frac{-1 - \sqrt{5}}{2}.$$

It remains to determine x by solving the two equations

$$x + \frac{1}{x} = y_1, \quad \text{and} \quad x + \frac{1}{x} = y_2$$

which are equivalent to

$$x^2 - y_1 x + 1 = 0, \quad \text{and} \quad x^2 - y_2 x + 1 = 0.$$

The four roots found by solving these are

$$x_1 = \frac{-1 + \sqrt{5}}{4} + i\frac{\sqrt{10 + 2\sqrt{5}}}{4},$$

$$x_2 = \frac{-1 + \sqrt{5}}{4} - i\frac{\sqrt{10 + 2\sqrt{5}}}{4},$$

$$x_3 = \frac{-1 - \sqrt{5}}{4} + i\frac{\sqrt{10 - 2\sqrt{5}}}{4},$$

$$x_4 = \frac{-1 - \sqrt{5}}{4} - i\frac{\sqrt{10 - 2\sqrt{5}}}{4}.$$

Method 2. Another approach to this problem is to multiply each side of the original equation by $x - 1$. Since

$$(x - 1)(x^4 + x^3 + x^2 + x + 1) = x^5 - 1$$

an equivalent problem is to find all x (other than $x = 1$) which satisfy

$$x^5 = 1.$$

These are the five fifth roots of unity, given by

$$x_1 = \cos\frac{2}{5}\pi + i\sin\frac{2}{5}\pi,$$

$$x_2 = \cos \frac{4}{5}\pi + i \sin \frac{4}{5}\pi,$$

$$x_3 = \cos \frac{6}{5}\pi + i \sin \frac{6}{5}\pi,$$

$$x_4 = \cos \frac{8}{5}\pi + i \sin \frac{8}{5}\pi,$$

$$x_5 = 1.$$

Problem 2.   Find the solution of the equation

$$\cos x + \cos 3x + \cdots + \cos(2n-1)x = 0 \tag{1}$$

where $n \in \mathbf{N}$.

Solution.   Obviously, (1) is invariant under

$$x \to x + \pi.$$

Now (1) can be written in the form

$$\Re(e^{ix} + e^{3ix} + \cdots + e^{(2n-1)ix}) = 0$$

where $\Re$ denotes the real part of a complex number. The sum on the left hand side is a geometric series. Therefore we obtain for the real part

$$\frac{\sin(2nx)}{2 \sin x} = 0.$$

Consequently,

$$x = \frac{\pi}{2n}, \frac{2\pi}{2n}, \frac{3\pi}{2n}, \ldots, \frac{(2n-1)\pi}{2n} \qquad \text{modulo } \pi.$$

Remark.   *The solution of this problem also gives the extrema of the function*

$$f(x) = \sin x + \frac{1}{3}\sin 3x + \cdots + \frac{1}{2n-1}\sin(2n-1)x.$$

The function f is a Fourier approximation.

Problem 3.   (i) Solve the equation

$$\sin((2n+1)\phi) = 0 \tag{1}$$

where $n \in \mathbf{Z}$ and $\phi \in \mathbf{R}$.

(ii) Solve the equation

$$\sin z = 0 \tag{2}$$

where $z \in \mathbf{C}$.

Solution.   (i) From

$$\sin x = 0$$

with $x \in \mathbf{R}$, we obtain

$$x = k\pi, \quad k \in \mathbf{Z}.$$

Therefore

$$(2n + 1)\phi = k\pi$$

or

$$\phi = \frac{k\pi}{2n + 1}.$$

(ii) Since

$$\sin z \equiv \sin(x + iy) \equiv \sin x \cos(iy) + \cos x \sin(iy) \equiv \sin x \cosh y + i \cos x \sinh y$$

we have

$$\sin x \cosh y + i \cos x \sinh y = 0.$$

It follows that

$$\sin x \cosh y = 0, \tag{3}$$

$$\cos x \sinh y = 0. \tag{4}$$

Since

$$\cosh y \neq 0$$

for $y \in \mathbf{R}$, Eq. (3) can only be satisfied if $x = k\pi$ where $k \in \mathbf{Z}$. Since

$$\cos(k\pi) \neq 0$$

if $k \in \mathbf{Z}$, Eq. (4) can only be satisfied if $y = 0$. Consequently, the solution to (2) is $z = x + iy$ with $x = k\pi$ and $y = 0$.

Problem 4. A real-valued function f, defined on the rational numbers $\mathbf{Q}$, satisfies

$$f(x + y) = f(x) + f(y)$$

for all rational x and y. Prove that

$$f(x) = f(1) \cdot x$$

for all rational x.

Solution. Let $n \in \mathbf{N}$. Then

$$f(1) = f \underbrace{\left(\frac{1}{n} + \cdots + \frac{1}{n} \right)}_{n}.$$

Therefore

$$f \left(\frac{1}{n} \right) = \frac{1}{n} f(1).$$

Any rational number can be represented as m/n with m an integer and $n \in \mathbf{N}$. Then

$$f \left(\frac{m}{n} \right) = f \underbrace{\left(\frac{1}{n} + \cdots + \frac{1}{n} \right)}_{m},$$

$$f \left(\frac{m}{n} \right) = f \underbrace{\left(\frac{1}{n} \right) + \cdots + f \left(\frac{1}{n} \right)}_{m}.$$

Thus

$$f \left(\frac{m}{n} \right) = m f \left(\frac{1}{n} \right) = \frac{m}{n} f(1).$$

Chapter 5

Matrix Calculations

Problem 1. (i) Let R be a *nonsingular* $n \times n$ matrix (i.e., R^{-1} exists). Let A and B be two arbitrary $n \times n$ matrices. Assume that $R^{-1}AR$ and $R^{-1}BR$ are diagonal matrices. Show that

$$[A, B] = 0$$

where $[A, B]$ denotes the *commutator*, i.e.

$$[A, B] := AB - BA.$$

(ii) Let X be an arbitrary $n \times n$ matrix. Let U be a nonsingular $n \times n$ matrix. Assume that $UXU^{-1} = X$. Show that $[X, U] = 0$.

Solution. (i) Since $R^{-1}AR$ and $R^{-1}BR$ are diagonal matrices it follows that

$$[R^{-1}AR, R^{-1}BR] = 0.$$

This means that $R^{-1}AR$ and $R^{-1}BR$ commute. Therefore

$$R^{-1}ARR^{-1}BR - R^{-1}BRR^{-1}AR = 0.$$

It follows that

$$R^{-1}ABR - R^{-1}BAR = R^{-1}[A, B]R = 0.$$

Finally

$$RR^{-1}[A, B]RR^{-1} = [A, B] = 0.$$

(ii) From $UXU^{-1} = X$, we obtain

$$UXU^{-1} - X = 0.$$

Multiplying from the right with U, we find

$$UX - XU \equiv [U, X] = 0.$$

Problem 2. Let A and B be two $n \times n$ matrices. Assume that B is nonsingular. This means that

$$\det B \neq 0$$

and the inverse of B exists. Show that

$$[A, B^{-1}] = -B^{-1}[A, B]B^{-1}$$

where $[,]$ denotes the commutator.

Solution. We have

$$[A, B^{-1}] \equiv AB^{-1} - B^{-1}A.$$

Since

$$BB^{-1} = I$$

and

$$B^{-1}B = I$$

where I is the $n \times n$ unit matrix, we can write

$$[A, B^{-1}] = AB^{-1} - B^{-1}A = B^{-1}BAB^{-1} - B^{-1}ABB^{-1}.$$

Thus

$$[A, B^{-1}] = -B^{-1}(AB - BA)B^{-1} = -B^{-1}[A, B]B^{-1}.$$

Problem 3.   Let $a, b, c \in \mathbf{R}$ and let

$$A(a,b,c) = \begin{pmatrix} a & b & 0 & 0 & \cdots & 0 & 0 & 0 \\ c & a & b & 0 & \cdots & 0 & 0 & 0 \\ 0 & c & a & b & \cdots & 0 & 0 & 0 \\ & & \ddots & & \ddots & & & \\ \vdots & & & \ddots & & \ddots & & \\ & & & & \ddots & & \ddots & \\ 0 & 0 & 0 & 0 & \cdots & c & a & b \\ 0 & 0 & 0 & 0 & \cdots & 0 & c & a \end{pmatrix}$$

be an $n \times n$ matrix. In other words

$$A_{jk} = \begin{cases} b & \text{if } j = k+1, \\ c & \text{if } j = k-1, \\ a & \text{if } j = k, \\ 0 & \text{otherwise} \end{cases}$$

with $j, k = 1, 2, \ldots, n$. Calculate

$$\det A(a,b,c).$$

Solution.   Let

$$D_n(a,b,c) := \det A(a,b,c).$$

By expanding in minors on the first row, we find the linear second-order difference equation with constant coefficients

$$D_{n+2}(a,b,c) = aD_{n+1}(a,b,c) - bcD_n(a,b,c) \tag{1}$$

where $n = 1, 2, \ldots$. Obviously the initial values of the difference Eq. (1) are given by

$$D_1(a,b,c) = a, \quad D_2(a,b,c) = a^2 - bc.$$

Since the linear difference Eq. (1) has constant coefficients we can solve it with the ansatz

$$D_n(a,b,c) = kr^n \tag{2}$$

where k is a constant, or we may employ the recurrence relations for the *Chebyshev polynomials* to obtain

$$D_n(a, b, c) = \begin{cases} (bc)^{n/2} U_n \left(\dfrac{a}{2\sqrt{bc}} \right) & \text{if } bc \neq 0, \\ a^n & \text{if } bc = 0 \end{cases}$$

where U_n is the n-th degree Chebyshev polynomial of the second kind given by

$$U_n(s) = \frac{(s + \sqrt{s^2 - 1})^{n+1} - (s - \sqrt{s^2 - 1})^{n+1}}{2\sqrt{s^2 - 1}}.$$

We notice that

$$a^n = \lim_{bc \to 0} (bc)^{n/2} U_n \left(\frac{a}{2\sqrt{bc}} \right).$$

The solution can also be given as

$$D_n(a, b, c) = a^n \prod_{j=1}^{n} \left(1 - \frac{2\sqrt{bc}}{a} \cos \left(\frac{j\pi}{n+1} \right) \right).$$

Remark. *The ansatz (2) leads to*

$$r^2 = ar - bc$$

with the solution

$$r_{1,2} = \frac{a}{2} \pm \sqrt{\frac{a^2}{4} - bc}.$$

Problem 4. (i) Let A, B be two $n \times n$ matrices. Assume that

$$\text{tr } A = 0, \quad \text{tr } B = 0$$

where tr denotes the *trace*. Let C be an $n \times n$ matrix. The trace of C is defined as

$$\text{tr } C := \sum_{j=1}^{n} c_{jj}.$$

Can we conclude that $\text{tr}(AB) = 0$?

(ii) Let A and B be two $n \times n$ matrices. Prove that

$$\text{tr}(AB) \neq (\text{tr}\, A)(\text{tr}\, B)$$

in general.

Solution.   (i) The answer is no. Let

$$A = B = \begin{pmatrix} 0 & 1 \\ 1 & 0 \end{pmatrix}.$$

Then

$$\text{tr}\, A = \text{tr}\, B = 0.$$

However, since

$$AB = \begin{pmatrix} 1 & 0 \\ 0 & 1 \end{pmatrix}$$

we have

$$\text{tr}(AB) = 2.$$

(ii) As a counterexample, we can consider the matrices given in (i).

Problem 5.   Let A be an $n \times n$ *hermitian matrix*, i.e.,

$$A = A^* \equiv \bar{A}^T$$

where $^-$ denotes the complex conjugate and ^{T} the transpose.

(i) Show that $A + iI$ is invertible, where I denotes the $n \times n$ unit matrix.
(ii) Show that

$$U := (A - iI)(A + iI)^{-1}$$

is a unitary matrix.

Solution.   (i) Since A is hermitian we can find a unitary matrix V such that VAV^* is a diagonal matrix, where $V^* = V^{-1}$. The diagonal elements are real (they are the eigenvalues of A). Thus

$$V(A+iI)V^* = VAV^*+iVIV^* = VAV^*+iI = \text{diag}(\lambda_1+i, \lambda_2+i, \ldots, \lambda_n+i)$$

The inverse of this diagonal matrix exists and therefore the inverse of $A+i$ exists since $\det(XYZ) \equiv \det(X)\det(Y)\det(Z)$ for $n \times n$ matrices X, Y, Z

(ii) Recall that U is a *unitary matrix* if

$$U^*U = I$$

where

$$U^* \equiv \bar{U}^T.$$

We note that $A - iI$ and $A + iI$ commute with each other, i.e.,

$$[A - iI, A + iI] = 0.$$

Now

$$U^* = [(A + iI)^{-1}]^*(A - iI)^* = [(A + iI)^*]^{-1}(A + iI) = (A - iI)^{-1}(A + iI).$$

Consequently,

$$U^*U = (A - iI)^{-1}(A + iI)(A - iI)(A + iI)^{-1}$$
$$= (A - iI)^{-1}(A - iI)(A + iI)(A + iI)^{-1} = I.$$

Thus U is a unitary matrix.

Remark.   *Let A be an $n \times n$ hermitian matrix. Then*

$$\exp(iA)$$

is a unitary matrix.

Problem 6.   Let X, Y and Z be arbitrary $n \times n$ matrices.

i) Show that

$$[X, Y + Z] = [X, Y] + [X, Z].$$

ii) Show that

$$[X, Y] = -[Y, X].$$

iii) Calculate

$$[X, [Y, Z]] + [Z, [X, Y]] + [Y, [Z, X]]$$

where $[,]$ denotes the commutator, i.e.,

$$[X, Y] := XY - YX.$$

Solution.   (i) Straightforward calculation yields

$$[X, Y + Z] = X(Y + Z) - (Y + Z)X$$
$$= XY - YX + XZ - ZX = [X, Y] + [X, Z].$$

(ii) Straightforward calculation yields

$$[X, Y] = XY - YX = -(YX - XY) = -[Y, X].$$

(iii) Since

$$[X, [Y, Z]] = [X, YZ - ZY] = XYZ - YZX - XZY + ZYX, \qquad (1)$$
$$[Z, [X, Y]] = [Z, XY - YX] = ZXY - XYZ - ZYX + YXZ,$$
$$[Y, [Z, X]] = [Y, ZX - XZ] = YZX - ZXY - YXZ + XZY, \qquad (2)$$

we obtain, by adding (1) through (2),

$$[X, [Y, Z]] + [Z, [X, Y]] + [Y, [Z, X]] = 0. \qquad (3)$$

Remark.   *Equation (3) is called the* Jacobi identity. *The $n \times n$ matrices over the real or complex numbers form a* Lie algebra *under the commutator. The $n \times n$ matrices over the real or complex numbers form an* associative algebra *with unit element (the unit matrix) under matrix multiplication.*

Problem 7.   Let A, B and H be three $n \times n$ matrices. Assume that

$$[A, H] = 0$$

and

$$[B, H] = 0.$$

Calculate

$$[[A, B], H].$$

Solution.   For arbitrary $n \times n$ matrices X, Y and Z we have (Jacobi identity)

$$[X, [Y, Z]] + [Z, [X, Y]] + [Y, [Z, X]] = 0.$$

Now let $X \equiv A$, $Y \equiv B$ and $Z \equiv H$. Since $[A, H] = 0$ and $[B, H] = 0$, it follows that

$$[H, [A, B]] = 0.$$

Since

$$[H, [A, B]] = -[[A, B], H],$$

we obtain

$$[[A, B], H] = 0.$$

Remark. *If we assume*

$$[H, A] = A, \quad [B, H] = B$$

we also find

$$[[A, B], H] = 0.$$

Problem 8. Let A and B be two $n \times n$ matrices. Assume that A^{-1} exists. Find the expansion of

$$(A - \epsilon B)^{-1}$$

as a power series in ϵ, where ϵ is a real parameter.

Hint. Set

$$(A - \epsilon B)^{-1} := \sum_{n=0}^{\infty} \epsilon^n L_n. \tag{1}$$

Solution. Multiplying (1) on the left by $A - \epsilon B$, we obtain

$$I = \sum_{n=0}^{\infty} \epsilon^n (A - \epsilon B) L_n = AL_0 + \sum_{n=1}^{\infty} \epsilon^n (AL_n - BL_{n-1})$$

where I is the $n \times n$ unit matrix. By equating coefficients of powers of ϵ, we find that

$$L_0 = A^{-1}$$

and

$$L_n = A^{-1} B L_{n-1}$$

where $n = 1, 2, \ldots$. Consequently,

$$(A - \epsilon B)^{-1} = A^{-1} + \epsilon A^{-1} B A^{-1} + \epsilon^2 A^{-1} B A^{-1} B A^{-1} + \cdots. \tag{2}$$

Remark.   *If $A = I$, then (2) takes the form*

$$(I - \epsilon B)^{-1} = I + \epsilon B + \epsilon^2 B^2 + \cdots$$

Discuss the radius of convergence of this series.

Problem 9.   (i) Let A and B be two hermitian matrices. Is the commutator $[A, B]$ again a hermitian matrix?

(ii) Let A and B be two skew-hermitian matrices. Is the commutator $[A, B]$ again a skew-hermitian matrix?

Solution.   (i) Since A and B are hermitian matrices we have

$$A^* = A, \quad B^* = B.$$

Now

$$([A, B])^* = (AB - BA)^* = (AB)^* - (BA)^* = B^* A^* - A^* B^*.$$

Thus

$$([A, B])^* = BA - AB = [B, A].$$

Consequently, in general the commutator of two hermitian matrices is not a hermitian matrix, since

$$[A, B] \neq [B, A]$$

in general.

(ii) Since A and B are two *skew-hermitian matrices* we have

$$A^* = -A, \quad B^* = -B.$$

Now

$$([A, B])^* = (AB - BA)^* = (AB)^* - (BA)^*.$$

Thus

$$([A, B])^* = B^* A^* - A^* B^* = BA - AB = -[A, B].$$

Consequently, the commutator of two skew-hermitian matrices is again a skew-hermitian matrix.

Remark.   *Notice that*

$$\text{tr}([X, Y]) = 0$$

for two arbitrary $n \times n$ matrices X and Y over $\mathbf{R}$ or $\mathbf{C}$.

Problem 10.   Let A, B and C be three arbitrary $n \times n$ matrices.

(i) Show that

$$\text{tr}(AB) = \text{tr}(BA). \tag{1}$$

(ii) Calculate

$$\text{tr}([A, B]).$$

(iii) Find 2×2 matrices A, B and C such that

$$\text{tr}(ABC) \neq \text{tr}(ACB).$$

Solution.   (i) Since

$$(AB)_{ij} = \sum_{k=1}^{n} a_{ik} b_{kj},$$

the diagonal elements of AB are given by

$$(AB)_{ii} = \sum_{k=1}^{n} a_{ik} b_{ki}.$$

Since

$$(BA)_{ij} = \sum_{k=1}^{n} b_{ik} a_{kj},$$

the diagonal elements of BA are given by

$$(BA)_{ii} = \sum_{k=1}^{n} b_{ik} a_{ki} = \sum_{k=1}^{n} a_{ki} b_{ik}.$$

Now

$$\text{tr}(AB) = (AB)_{11} + (AB)_{22} + \cdots + (AB)_{nn}$$

and

$$\text{tr}(BA) = (BA)_{11} + (BA)_{22} + \cdots + (BA)_{nn},$$

so that

$$\text{tr}(AB) = \sum_{i=1}^{n} \sum_{k=1}^{n} a_{ik} b_{ki},$$

$$\text{tr}(BA) = \sum_{i=1}^{n} \sum_{k=1}^{n} a_{ki} b_{ik}.$$

Therefore, identity (1) holds.

(ii) Using identity (1), we have

$$\text{tr}([A, B]) = \text{tr}(AB - BA) = \text{tr}(AB) - \text{tr}(BA) = 0.$$

(iii) Let

$$A = \begin{pmatrix} 1 & 0 \\ 0 & 0 \end{pmatrix}, \quad B = \begin{pmatrix} 0 & 1 \\ 0 & 0 \end{pmatrix}, \quad C = \begin{pmatrix} 0 & 0 \\ 1 & 0 \end{pmatrix}.$$

Then

$$ABC = \begin{pmatrix} 1 & 0 \\ 0 & 0 \end{pmatrix}$$

and

$$ACB = \begin{pmatrix} 0 & 0 \\ 0 & 0 \end{pmatrix}.$$

Consequently

$$\text{tr}(ABC) = 1$$

and

$$\text{tr}(ACB) = 0.$$

Problem 11.   Let $A(\epsilon)$ be an $n \times n$ matrix which depends on a real parameter ϵ. Assume that $dA/d\epsilon$ and $A^{-1}(\epsilon)$ exist for all ϵ. Show that

$$\frac{dA^{-1}}{d\epsilon} = -A^{-1}(\epsilon) \frac{dA}{d\epsilon} A^{-1}(\epsilon). \tag{1}$$

Solution.   We start from the identity

$$A(\epsilon) A^{-1}(\epsilon) = I \tag{2}$$

where I is the $n \times n$ unit matrix. Taking the derivative of (2) with respect to ϵ yields

$$\frac{dA}{d\epsilon} A^{-1}(\epsilon) + A(\epsilon)\frac{dA^{-1}}{d\epsilon} = 0$$

since

$$\frac{dI}{d\epsilon} = 0 .$$

Therefore

$$\frac{dA}{d\epsilon} A^{-1}(\epsilon) = -A(\epsilon)\frac{dA^{-1}}{d\epsilon} .$$

Multiplying both sides from the left by $A^{-1}(\epsilon)$, we find (1).

Problem 12. Let A and L be two $n \times n$ matrices. Calculate

$$e^L A e^{-L} .$$

Solution. We consider

$$A(\epsilon) := e^{\epsilon L} A e^{-\epsilon L}$$

where ϵ is a real parameter. Taking the derivative of $A(\epsilon)$ with respect to ϵ yields

$$\frac{dA}{d\epsilon} = L e^{\epsilon L} A e^{-\epsilon L} - e^{\epsilon L} A e^{-\epsilon L} L = [L, A(\epsilon)] .$$

The second derivative gives

$$\frac{d^2 A}{d\epsilon^2} = \left[L, \frac{dA}{d\epsilon} \right] = [L, [L, A(\epsilon)]]$$

and so on. Consequently, we can write the matrix $e^L A e^{-L} = A(1)$ as a Taylor series expansion about the origin, i.e.,

$$A(1) = A(0) + \frac{1}{1!}\frac{dA(0)}{d\epsilon} + \frac{1}{2!}\frac{d^2 A(0)}{d\epsilon^2} + \cdots$$

where $A(0) \equiv A$ and

$$\frac{dA(0)}{d\epsilon} \equiv \frac{dA(\epsilon)}{d\epsilon}\bigg|_{\epsilon=0} .$$

Thus we find

$$e^L A e^{-L} \equiv A + [L, A] + \frac{1}{2!}[L, [L, A]] + \frac{1}{3!}[L, [L, [L, A]]] + \cdots .$$

Problem 13.   Let A and B be two $n \times n$ matrices. Assume that

$$[[A, B], A] = 0, \quad [[A, B], B] = 0. \tag{1}$$

Show that

$$e^A e^B \equiv e^{A+B+1/2[A,B]}. \tag{2}$$

Solution.   Let

$$T(\epsilon) := e^{\epsilon A} e^{\epsilon B} \tag{3}$$

where ϵ is a real parameter. Differentiating (3) with respect to ϵ gives

$$\frac{dT}{d\epsilon} = A e^{\epsilon A} e^{\epsilon B} + e^{\epsilon A} B e^{\epsilon B} = (A + e^{\epsilon A} B e^{-\epsilon A}) T(\epsilon).$$

Using (1), we find

$$e^{\epsilon A} B e^{-\epsilon A} = B - [B, A]\epsilon.$$

Then

$$\frac{dT}{d\epsilon} = (A + B + [A, B]\epsilon) T(\epsilon). \tag{4}$$

Consequently, T is the solution to the matrix differential Eq. (4) with the initial condition

$$T(\epsilon = 0) = I$$

where I is the $n \times n$ unit matrix. The initial condition follows from (3). Since the matrices $A + B$ and $[A, B]$ commute, the matrix differential Eq. (4) can be integrated as if the matrices were merely numbers, to give the solution

$$T(\epsilon) = e^{\epsilon(A+B)} e^{1/2\epsilon^2[A,B]}.$$

The identity (2) follows by setting $\epsilon = 1$.

Problem 14.   Let $\sigma := (\sigma_1, \sigma_2, \sigma_3)$ where

$$\sigma_1 := \begin{pmatrix} 0 & 1 \\ 1 & 0 \end{pmatrix}, \quad \sigma_2 := \begin{pmatrix} 0 & -i \\ i & 0 \end{pmatrix}, \quad \sigma_3 := \begin{pmatrix} 1 & 0 \\ 0 & -1 \end{pmatrix}.$$

Let $\mathbf{a} := (a_1, a_2, a_3)$ where $a_j \in \mathbf{R}$ and

$$\mathbf{a} \cdot \sigma := a_1\sigma_1 + a_2\sigma_2 + a_3\sigma_3 \,.$$

i) Calculate $(\mathbf{a} \cdot \sigma)^2$, $(\mathbf{a} \cdot \sigma)^3$ and $(\mathbf{a} \cdot \sigma)^4$.

ii) Calculate

$$\exp(\mathbf{a} \cdot \sigma) \,.$$

Remark. σ_1, σ_2 and σ_3 *are the so-called* Pauli matrices.

Solution. (i) First, we notice that

$$\sigma_1^2 = I \,, \quad \sigma_2^2 = I \,, \quad \sigma_3^2 = I$$

where I is the 2×2 unit matrix. Furthermore

$$\sigma_1\sigma_2 + \sigma_2\sigma_1 = 0 \,, \quad \sigma_1\sigma_3 + \sigma_3\sigma_1 = 0 \,, \quad \sigma_2\sigma_3 + \sigma_3\sigma_2 = 0$$

where 0 is the 2×2 zero matrix. Thus

$$(\mathbf{a} \cdot \sigma)^2 = (a_1\sigma_1 + a_2\sigma_2 + a_3\sigma_3)(a_1\sigma_1 + a_2\sigma_2 + a_3\sigma_3) = (a_1^2 + a_2^2 + a_3^2)I \,.$$

In the following we set

$$a^2 \equiv \mathbf{a}^2 := a_1^2 + a_2^2 + a_3^2 \,.$$

It follows that

$$(\mathbf{a} \cdot \sigma)^3 = (\mathbf{a} \cdot \sigma)^2(\mathbf{a} \cdot \sigma) = (a^2 I)(\mathbf{a} \cdot \sigma) = a^2(\mathbf{a} \cdot \sigma)$$

and

$$(\mathbf{a} \cdot \sigma)^4 = (\mathbf{a} \cdot \sigma)^2(\mathbf{a} \cdot \sigma)^2 = (a^2 I)(a^2 I) = a^4 I \,.$$

ii) From the definition

$$e^{\mathbf{a} \cdot \sigma} := \sum_{k=0}^{\infty} \frac{(\mathbf{a} \cdot \sigma)^k}{k!}$$

and the result from (i), we have

$$e^{\mathbf{a} \cdot \sigma} = I + \mathbf{a} \cdot \sigma + \frac{(\mathbf{a} \cdot \sigma)^2}{2!} + \frac{(\mathbf{a} \cdot \sigma)^3}{3!} + \frac{(\mathbf{a} \cdot \sigma)^4}{4!} + \frac{(\mathbf{a} \cdot \sigma)^5}{5!} + \cdots$$

$$= I + \mathbf{a} \cdot \sigma + \frac{a^2 I}{2!} + \frac{a^2 (\mathbf{a} \cdot \sigma)}{3!} + \frac{a^4 I}{4!} + \frac{a^4 (\mathbf{a} \cdot \sigma)}{5!} + \cdots$$

$$= \left(1 + \frac{a^2}{2!} + \frac{a^4}{4!} + \cdots \right) I + \left(1 + \frac{a^2}{3!} + \frac{a^4}{5!} + \cdots \right) \mathbf{a} \cdot \sigma.$$

Thus

$$e^{\mathbf{a} \cdot \sigma} = (\cosh a) I + \frac{1}{a} (\sinh a)(\mathbf{a} \cdot \sigma)$$

for $a \neq 0$. If $\mathbf{a} = (0, 0, 0)$, we have

$$e^{\mathbf{a} \cdot \sigma} = \begin{pmatrix} 1 & 0 \\ 0 & 1 \end{pmatrix}.$$

Problem 15.   Let A be an $n \times n$ matrix. The entries depend smoothly on a real parameter ϵ. Show that in general

$$A(\epsilon) \frac{dA}{d\epsilon} \neq \frac{dA}{d\epsilon} A(\epsilon). \tag{1}$$

Solution.   Let

$$A(\epsilon) := \begin{pmatrix} \epsilon & \epsilon^2 \\ \epsilon^3 & 1 \end{pmatrix}.$$

Then

$$\frac{dA}{d\epsilon} = \begin{pmatrix} 1 & 2\epsilon \\ 3\epsilon^2 & 0 \end{pmatrix}.$$

Thus

$$A \frac{dA}{d\epsilon} = \begin{pmatrix} \epsilon + 3\epsilon^4 & 2\epsilon^2 \\ 3\epsilon^2 + \epsilon^3 & 2\epsilon^4 \end{pmatrix}$$

and

$$\frac{dA}{d\epsilon} A = \begin{pmatrix} \epsilon + 2\epsilon^4 & 2\epsilon + \epsilon^2 \\ 3\epsilon^3 & 3\epsilon^4 \end{pmatrix}.$$

Thus (1) holds in general.

Remark.   *In particular, we have*

$$\frac{d}{d\epsilon}A^2 = A\frac{dA}{d\epsilon} + \frac{dA}{d\epsilon}A \neq 2A\frac{dA}{d\epsilon}$$

in general.

Problem 16.   Let A be a square finite-dimensional matrix over $\mathbf{R}$ such that

$$AA^T = I. \tag{1}$$

(i) Show that

$$A^T A = I. \tag{2}$$

(ii) Does (2) also hold for infinite dimensional matrices?

Solution.   (i) Since $\det A = \det A^T$ and $\det I = 1$, we obtain from (1) that

$$(\det A)^2 = 1.$$

Therefore the inverse of A exists and we have $A^T = A^{-1}$ with $A^{-1}A = AA^{-1} = I$.

(ii) The answer is no. Let

$$A = \begin{pmatrix} 0 & 1 & 0 & 0 & 0 & \cdots \\ 0 & 0 & 1 & 0 & 0 & \cdots \\ 0 & 0 & 0 & 1 & 0 & \cdots \\ \vdots & \vdots & \vdots & \vdots & \ddots & \cdots \end{pmatrix}.$$

Then the transpose matrix A^T of A is given by

$$A^T = \begin{pmatrix} 0 & 0 & 0 & 0 & 0 & \cdots \\ 1 & 0 & 0 & 0 & 0 & \cdots \\ 0 & 1 & 0 & 0 & 0 & \cdots \\ 0 & 0 & 1 & 0 & 0 & \cdots \\ \vdots & \vdots & \vdots & \vdots & \ddots & \cdots \end{pmatrix}.$$

It follows that

$$AA^T = \mathrm{diag}(1, 1, 1, \ldots) \equiv I$$

and

$$A^T A = \operatorname{diag}(0, 1, 1, \ldots)$$

where I is the infinite-dimensional unit matrix. Consequently,

$$A^T A \neq A A^T.$$

Problem 17.   Let

$$\hat{H} = \hbar \omega \sigma_z$$

be a Hamilton operator acting in the two-dimensional Hilbert space $\mathbb{C}^2$, where

$$\sigma_z := \begin{pmatrix} 1 & 0 \\ 0 & -1 \end{pmatrix}$$

and ω is the frequency. Calculate the time evolution of

$$\sigma_x := \begin{pmatrix} 0 & 1 \\ 1 & 0 \end{pmatrix}.$$

Remark.   *The matrices σ_x, σ_y and σ_z are the Pauli matrices, where*

$$\sigma_y := \begin{pmatrix} 0 & -i \\ i & 0 \end{pmatrix}.$$

The Pauli matrices form a Lie algebra under the commutator.

Solution.   The *Heisenberg equation of motion* is given by

$$i\hbar \frac{d\sigma_x}{dt} = [\sigma_x, \hat{H}](t).$$

Since

$$[\sigma_x, \hat{H}] = \hbar \omega [\sigma_x, \sigma_z] = -2i\hbar \omega \sigma_y$$

we obtain

$$\frac{d\sigma_x}{dt} = -2\omega \sigma_y(t).$$

Now we have to calculate the time-evolution of σ_y, i.e.,

$$i\hbar \frac{d\sigma_y}{dt} = [\sigma_y, \hat{H}](t).$$

Since

$$[\sigma_y, \hat{H}] = \hbar\omega[\sigma_y, \sigma_z] = 2i\hbar\omega\sigma_x \,,$$

we find

$$\frac{d\sigma_y}{dt} = 2\omega\sigma_x(t) \,.$$

To summarize, we have to solve the following system of linear matrix differential equations with constant coefficients

$$\frac{d\sigma_x}{dt} = -2\omega\sigma_y(t) \,, \tag{1a}$$

$$\frac{d\sigma_y}{dt} = 2\omega\sigma_x(t) \,. \tag{1b}$$

The initial conditions of system (11) are

$$\sigma_x(t=0) = \sigma_x \,, \quad \sigma_y(t=0) = \sigma_y \,.$$

Then the solution of the initial value problem is given by

$$\sigma_x(t) = \sigma_x \cos(2\omega t) - \sigma_y \sin(2\omega t) \,,$$

$$\sigma_y(t) = \sigma_y \cos(2\omega t) + \sigma_x \sin(2\omega t) \,.$$

Remark.   *The solution of the Heisenberg equation of motion can also be given as*

$$\sigma_x(t) = e^{i\hat{H}t/\hbar}\sigma_x e^{-i\hat{H}t/\hbar} \,,$$

$$\sigma_y(t) = e^{i\hat{H}t/\hbar}\sigma_y e^{-i\hat{H}t/\hbar} \,.$$

Problem 18.   Let

$$\mathbf{a}_1 = \frac{a}{2}(1, 1, -1) \,,$$

$$\mathbf{a}_2 = \frac{a}{2}(-1, 1, 1) \,,$$

$$\mathbf{a}_3 = \frac{a}{2}(1, -1, 1) \,,$$

where a is a positive constant. Find the vectors $\mathbf{b}_j$ ($j = 1, 2, 3$) such that

$$\mathbf{b}_j \cdot \mathbf{a}_k = 2\pi\delta_{jk} \tag{1}$$

where $\mathbf{b}_j \cdot \mathbf{a}_k$ denotes the scalar product of $\mathbf{b}_j$ and $\mathbf{a}_k$, i.e.,

$$\mathbf{b}_j \cdot \mathbf{a}_k := \sum_{n=1}^{3} b_{jn} a_{kn}$$

and

$$\delta_{jk} := \begin{cases} 1 & j = k \\ 0 & \text{otherwise} \end{cases}.$$

The vectors $\{\mathbf{a}_j : j = 1, 2, 3\}$ are a basis of the *body-centered cubic lattice*. The vectors $\{\mathbf{b}_j : j = 1, 2, 3\}$ are called the *reciprocal basis*.

Solution. In matrix notation (2) can be written as

$$BA = 2\pi I$$

where B is the matrix whose rows are the vectors $\mathbf{b}_j$, A is the matrix whose columns are the vectors $\mathbf{a}_j$ and I is the 3×3 unit matrix. Therefore

$$A = \frac{a}{2} \begin{pmatrix} 1 & -1 & 1 \\ 1 & 1 & -1 \\ -1 & 1 & 1 \end{pmatrix}.$$

We find

$$\det A \neq 0.$$

Consequently

$$B = 2\pi A^{-1}.$$

Using Gaussian elimination we find

$$B = \frac{2\pi}{a} \begin{pmatrix} 1 & 1 & 0 \\ 0 & 1 & 1 \\ 1 & 0 & 1 \end{pmatrix}.$$

Finally we arrive at

$$\mathbf{b}_1 = \frac{2\pi}{a}(1, 1, 0), \quad \mathbf{b}_2 = \frac{2\pi}{a}(0, 1, 1), \quad \mathbf{b}_3 = \frac{2\pi}{a}(1, 0, 1).$$

Remark.   *The vectors*

$$\frac{a}{2}(1,1,0),$$

$$\frac{a}{2}(0,1,1),$$

$$\frac{a}{2}(1,0,1)$$

are a basis of the face-centered cubic lattice.

Problem 19.   Let A be a linear transformation of a vector space V into itself. Suppose $\mathbf{x} \in V$ is such that

$$A^n\mathbf{x} = \mathbf{0}, \quad A^{n-1}\mathbf{x} \neq \mathbf{0}$$

for some positive integer n. Show that

$$\mathbf{x}, A\mathbf{x}, \dots, A^{n-1}\mathbf{x}$$

are linearly independent.

Solution.   Suppose that there are scalars $c_0, c_1, \dots, c_{n-1}$ such that

$$c_0\mathbf{x} + c_1 A\mathbf{x} + \dots + c_k A^k\mathbf{x} + \dots + c_{n-1}A^{n-1}\mathbf{x} = \mathbf{0}.$$

Applying A^{n-1} to both sides, we obtain

$$c_0 A^{n-1}\mathbf{x} + c_1 A^n\mathbf{x} + \dots + c_k A^{n-1+k}\mathbf{x} + \dots + c_{n-1}A^{n-1+n-1}\mathbf{x} = \mathbf{0}$$

where we used

$$A\mathbf{0} = \mathbf{0}.$$

Thus

$$c_0 A^{n-1}\mathbf{x} = \mathbf{0}$$

and therefore

$$c_0 = 0.$$

By the *induction principle* (multiplying by A^{n-k-1}) we find that all $c_k = 0$. Thus the set is linearly independent.

Problem 20.   Consider the boundary value problem (Dirichlet boundary conditions)

$$\frac{d^2u}{dx^2} + u = 0, \quad u(0) = u(1) = 1$$

for the interval $[0, 1]$. The exact solution is given by

$$u(x) = \cos(x) + \frac{1 - \cos(1)}{\sin(1)} \sin(x).$$

Find a matrix equation for the discretization

$$\frac{d^2u}{dx^2} \longrightarrow \frac{u_{i-1} - 2u_i + u_{i+1}}{h^2}, \quad u \longrightarrow u_i$$

where $h = 0.1$ and $u_i = u(i \cdot h)$.

Solution.   Since the interval is $[0, 1]$ and $h = 0.1$, we have $u_0, u_1, \ldots, u_{10}$ where u_0 and u_{10} are given by the boundary conditions. We have

$$
\begin{aligned}
i = 1: &\quad u_0 - 2u_1 + u_2 + h^2 u_1 = 0, \\
i = 2: &\quad u_1 - 2u_2 + u_3 + h^2 u_2 = 0, \\
i = 3: &\quad u_2 - 2u_3 + u_4 + h^2 u_3 = 0, \\
&\quad \cdots \quad \cdots \cdots \cdots \cdots \cdots \cdots \\
i = 9: &\quad u_8 - 2u_9 + u_{10} + h^2 u_9 = 0.
\end{aligned}
$$

The first equation can be brought into the form

$$-2u_1 + u_2 + h^2 u_1 = -u_0.$$

The last equation can be brought into the form

$$u_8 - 2u_9 + h^2 u_9 = -u_{10}.$$

From the boundary conditions $u(0) = 1$ and $u(1) = 1$, we obtain

$$u_0 = 1, \quad u_{10} = 1.$$

Thus we get the matrix form of the discretization with a 9×9 matrix

$$
\begin{pmatrix}
\alpha & 1 & 0 & 0 & 0 & 0 & 0 & 0 & 0 \\
1 & \alpha & 1 & 0 & 0 & 0 & 0 & 0 & 0 \\
0 & 1 & \alpha & 1 & 0 & 0 & 0 & 0 & 0 \\
0 & 0 & 1 & \alpha & 1 & 0 & 0 & 0 & 0 \\
0 & 0 & 0 & 1 & \alpha & 1 & 0 & 0 & 0 \\
0 & 0 & 0 & 0 & 1 & \alpha & 1 & 0 & 0 \\
0 & 0 & 0 & 0 & 0 & 1 & \alpha & 1 & 0 \\
0 & 0 & 0 & 0 & 0 & 0 & 1 & \alpha & 1 \\
0 & 0 & 0 & 0 & 0 & 0 & 0 & 1 & \alpha
\end{pmatrix}
\begin{pmatrix}
u_1 \\ u_2 \\ u_3 \\ u_4 \\ u_5 \\ u_6 \\ u_7 \\ u_8 \\ u_9
\end{pmatrix}
=
\begin{pmatrix}
-1 \\ 0 \\ 0 \\ 0 \\ 0 \\ 0 \\ 0 \\ 0 \\ -1
\end{pmatrix}
\qquad (1)
$$

with the abbreviation

$$
\alpha := -2 + h^2.
$$

The matrix on the left hand side of (1) is a tridiagonal matrix. The inverse of the matrix exists. The solution algorithm for the tridiagonal equation is called the tridiagonal solution (a variant of Gaussian elimination).

Chapter 6

Matrices and Groups

Problem 1. Let

$$A = \begin{pmatrix} 1 & 0 & 0 \\ 0 & 1 & 0 \\ 0 & 0 & 1 \end{pmatrix}, \quad B = \begin{pmatrix} 1 & 0 & 0 \\ 0 & 0 & 1 \\ 0 & 1 & 0 \end{pmatrix}, \quad C = \begin{pmatrix} 0 & 1 & 0 \\ 1 & 0 & 0 \\ 0 & 0 & 1 \end{pmatrix},$$

$$D = \begin{pmatrix} 0 & 1 & 0 \\ 0 & 0 & 1 \\ 1 & 0 & 0 \end{pmatrix}, \quad E = \begin{pmatrix} 0 & 0 & 1 \\ 1 & 0 & 0 \\ 0 & 1 & 0 \end{pmatrix}, \quad F = \begin{pmatrix} 0 & 0 & 1 \\ 0 & 1 & 0 \\ 1 & 0 & 0 \end{pmatrix}.$$

(i) Show that these matrices form a group under matrix multiplication.
(ii) Find all subgroups.

Remark. *These matrices are called* permutation matrices. *Why?*

Solution. (i) Since

$$AA = A, \quad AB = B, \quad AC = C, \quad AD = D, \quad AE = E, \quad AF = F,$$
$$BA = B, \quad BB = A, \quad BC = D, \quad BD = C, \quad BE = F, \quad BF = E,$$
$$CA = C, \quad CB = E, \quad CC = A, \quad CD = F, \quad CE = B, \quad CF = D,$$
$$DA = D, \quad DB = F, \quad DC = B, \quad DD = E, \quad DE = A, \quad DF = C,$$
$$EA = E, \quad EB = C, \quad EC = F, \quad ED = A, \quad EE = D, \quad EF = B,$$
$$FA = F, \quad FB = D, \quad FC = E, \quad FD = B, \quad FE = C, \quad FF = A,$$

we find the following: the set of matrices given above is closed under matrix multiplication; the neutral element is the matrix A, i.e. the unit matrix; each element has an inverse. From the table given above we also find that

$$A^{-1} = A, \quad B^{-1} = B, \quad C^{-1} = C, \quad D^{-1} = E, \quad E^{-1} = D, \quad F^{-1} = F.$$

Since the associative law holds for matrices, we find that the matrices given above form a finite group under matrix multiplication.

ii) The *order of a finite group* is the number of elements of the group. Thus our group has order 6. *Lagrange's theorem* tells us that the order of a subgroup of a finite group divides the order of the group. Thus the subgroups must have order 3, 2, 1. From the group table we find (besides the group itself) the subgroups

$$\{A, D, E\}$$

$$\{A, B\}, \{A, C\}, \{A, F\}$$

$$\{A\}$$

Remark 1. *The set of all $n \times n$ permutation matrices form a group under matrix multiplication.*

Remark 2. *Cayley's theorem tells us that every finite group is isomorphic to a subgroup (or the group itself) of these permutation matrices.*

Remark 3. *The order of an element $g \in G$ is the order of the cyclic subgroup generated by $\{g\}$, i.e. the smallest positive integer m such that*

$$g^m = e$$

here e is the identity element of the group. The integer m divides the order of G. Consider, for example, the element D of our group. Then

$$D^2 = E, \quad D^3 = A, \quad A \text{ identity element}.$$

Thus $m = 3$.

Problem 2.   (i) Let

$$A(\alpha) := \begin{pmatrix} \cos \alpha & \sin \alpha \\ -\sin \alpha & \cos \alpha \end{pmatrix}$$

here $\alpha \in \mathbf{R}$. Show that the matrices $A(\alpha)$ form a *group* under matrix multiplication.

(ii) Let

$$B(\alpha) := \begin{pmatrix} \cosh \alpha & \sinh \alpha \\ \sinh \alpha & \cosh \alpha \end{pmatrix}.$$

Show that the matrices $B(\alpha)$ form a group under matrix multiplication.

(iii) Let

$$X := \frac{dA(\alpha)}{d\alpha}\bigg|_{\alpha=0}.$$

Find X. Calculate $e^{\alpha X}$.

Solution.   (i) Since

$$A(\alpha)A(\beta) = \begin{pmatrix} \cos(\alpha+\beta) & \sin(\alpha+\beta) \\ -\sin(\alpha+\beta) & \cos(\alpha+\beta) \end{pmatrix}, \tag{1}$$

the set is closed under multiplication. To find (1) we have used the identities

$$\cos \alpha \cos \beta - \sin \alpha \sin \beta \equiv \cos(\alpha+\beta),$$

$$\sin \alpha \cos \beta + \cos \alpha \sin \beta \equiv \sin(\alpha+\beta).$$

For $\alpha = 0$ we obtain the neutral element of the group (i.e. the unit matrix

$$A(\alpha = 0) = \begin{pmatrix} 1 & 0 \\ 0 & 1 \end{pmatrix}.$$

Since $\det A(\alpha) = 1$, the inverse exists and is given by

$$A^{-1}(\alpha) = A(-\alpha) = \begin{pmatrix} \cos \alpha & -\sin \alpha \\ \sin \alpha & \cos \alpha \end{pmatrix}.$$

For arbitrary $n \times n$ matrices the associative law holds. Consequently the matrices $A(\alpha)$ form a group. The group is called SO(2).

(ii) Since

$$B(\alpha)B(\beta) = \begin{pmatrix} \cosh(\alpha+\beta) & \sinh(\alpha+\beta) \\ \sinh(\alpha+\beta) & \cosh(\alpha+\beta) \end{pmatrix}, \tag{2}$$

the set is closed under multiplication. To find (2) we have used the identities

$$\cosh \alpha \cosh \beta + \sinh \alpha \sinh \beta \equiv \cosh(\alpha+\beta),$$

$$\sinh \alpha \cosh \beta + \cosh \alpha \sinh \beta \equiv \sinh(\alpha+\beta).$$

For $\alpha = 0$ we obtain the neutral element

$$B(\alpha = 0) = \begin{pmatrix} 1 & 0 \\ 0 & 1 \end{pmatrix} .$$

Since $\det B(\alpha) = 1$, the inverse matrix exists and is given by

$$B^{-1}(\alpha) = B(-\alpha) = \begin{pmatrix} \cosh \alpha & -\sinh \alpha \\ -\sinh \alpha & \cosh \alpha \end{pmatrix} .$$

For arbitrary $n \times n$ matrices the associative law holds. Consequently, the matrices $B(\alpha)$ form a group. The group is called SO(1, 1).

iii) We find

$$X = \begin{pmatrix} 0 & 1 \\ -1 & 0 \end{pmatrix}$$

owing to $\sin(0) = 0$ and $\cos(0) = 1$. Since

$$e^{\alpha X} := \sum_{k=0}^{\infty} \frac{(\alpha X)^k}{k!} ,$$

we find

$$e^{\alpha X} = A(\alpha)$$

because $X^2 = -I$. In physics X is called the *generator* of SO(2).

Chapter 7

Matrices and Eigenvalue Problems

Problem 1.   (i) Show that the eigenvalues of a hermitian matrix are real.

(ii) Show that the eigenvalues of a skew-hermitian matrix are purely imaginary or zero.

(iii) Show that the eigenvalues λ_j of a unitary matrix satisfy $|\lambda_j| = 1$.

Solution.   (i) Since A is hermitian we have

$$A^* = A \tag{1}$$

where $A^* \equiv \bar{A}^T$. The *eigenvalue equation* is

$$A\mathbf{x} = \lambda\mathbf{x} \tag{2}$$

with $\mathbf{x} \neq \mathbf{0}$. Now we have the identity

$$(A\mathbf{x})^*\mathbf{x} \equiv \mathbf{x}^* A^*\mathbf{x} \equiv \mathbf{x}^*(A^*\mathbf{x}). \tag{3}$$

Inserting (1) and (2) into (3) gives

$$(\lambda\mathbf{x})^*\mathbf{x} = \mathbf{x}^*(\lambda\mathbf{x}).$$

Consequently

$$\bar{\lambda}(\mathbf{x}^*\mathbf{x}) = \lambda(\mathbf{x}^*\mathbf{x}).$$

Since $\mathbf{x}^*\mathbf{x} \neq 0$, we have $\bar{\lambda} = \lambda$. Therefore λ must be real.

(ii) Since A is skew-hermitian we have

$$A^* = -A. \tag{4}$$

The eigenvalue equation is

$$Ax = \lambda x. \tag{5}$$

Now using (4) we have

$$(Ax)^*x \equiv x^*A^*x \equiv x^*(A^*x) \equiv -x^*(Ax). \tag{6}$$

Inserting (5) into (6) yields

$$(\lambda x)^*x = -x^*(\lambda x)$$

or

$$\bar{\lambda}(x^*x) = -\lambda(x^*x).$$

Since $x^*x \neq 0$ we have

$$\bar{\lambda} = -\lambda.$$

Thus the eigenvalues are purely imaginary or zero.

iii) Since U is a unitary matrix we have

$$U^* = U^{-1}$$

where U^{-1} is the inverse of U. Let

$$Ux = \lambda x \tag{7}$$

be the eigenvalue equation. It follows that

$$(Ux)^* = (\lambda x)^*$$

or

$$x^*U^* = x^*\bar{\lambda}. \tag{8}$$

From (8) and (7) we find

$$x^*U^*Ux = \bar{\lambda}\lambda x^*x.$$

Since $U^*U = I$ we have

$$x^*x = \bar{\lambda}\lambda x^*x.$$

Since $x^*x \neq 0$ we have

$$\bar{\lambda}\lambda = 1.$$

Thus λ can be written as

$$\lambda = e^{i\alpha}, \alpha \in \mathbf{R}.$$

Problem 2.   Let

$$A = \begin{pmatrix} 0 & i \\ -i & 0 \end{pmatrix}.$$

(i) Calculate the eigenvalues and eigenvectors of the matrix A.
(ii) Are the eigenvectors orthogonal?
(iii) Do the eigenvectors form a basis in $\mathbf{C}^2$?
(iv) Find a unitary matrix U such that

$$U^* A U = \begin{pmatrix} \lambda_1 & 0 \\ 0 & \lambda_2 \end{pmatrix}.$$

Obviously, λ_1 and λ_2 are the eigenvalues of A.

Solution.   (i) The *eigenvalues* are determined by the equation

$$\det(A - \lambda I) = 0 \tag{1}$$

where det denotes the determinant and I is the 2×2 unit matrix. Equation (1) leads to

$$\lambda^2 - 1 = 0.$$

Thus we obtain the eigenvalues

$$\lambda_1 = 1, \quad \lambda_2 = -1.$$

Notice that the matrix A is hermitian and thus the eigenvalues must be real. The eigenvectors are determined by solving the equation

$$A \begin{pmatrix} x_1 \\ x_2 \end{pmatrix} = \lambda \begin{pmatrix} x_1 \\ x_2 \end{pmatrix}.$$

Therefore the normalized eigenvector which belongs to λ_1 is given by

$$\frac{1}{\sqrt{2}} \begin{pmatrix} 1 \\ -i \end{pmatrix}. \tag{2}$$

The normalized eigenvector which belongs to λ_2 is given by

$$\frac{1}{\sqrt{2}} \begin{pmatrix} 1 \\ i \end{pmatrix}. \tag{3}$$

ii) Two vectors **x** and **y** are called *orthogonal* if

$$\bar{\mathbf{x}}^T \mathbf{y} = 0.$$

The eigenvectors (2) and (3) are orthogonal, i.e.

$$\frac{1}{\sqrt{2}}(1, i)\frac{1}{\sqrt{2}} \begin{pmatrix} 1 \\ i \end{pmatrix} = 0.$$

iii) They form a basis in $\mathbf{C}^2$.

iv) The normalized eigenvectors given by (2) and (3) lead to the unitary matrix

$$U = \frac{1}{\sqrt{2}} \begin{pmatrix} 1 & 1 \\ -i & i \end{pmatrix}.$$

Consequently,

$$U^* = \frac{1}{\sqrt{2}} \begin{pmatrix} 1 & i \\ 1 & -i \end{pmatrix}$$

where $U^* = U^{-1}$. Then we obtain

$$U^* A U = \begin{pmatrix} 1 & 0 \\ 0 & -1 \end{pmatrix}.$$

Remark   *Let A be an $n \times n$ hermitian matrix. Assume that all eigenvalues of A are different. Then the eigenvectors are pairwise orthogonal.*

Problem 3.   Calculate the eigenvalues λ_j of the matrices

$$A = \begin{pmatrix} 0 & 0 & 1 \\ 0 & 1 & 0 \\ 1 & 0 & 0 \end{pmatrix}$$

and

$$B = \begin{pmatrix} 0 & 0 & 0 & 1 \\ 0 & 0 & 1 & 0 \\ 0 & 1 & 0 & 0 \\ 1 & 0 & 0 & 0 \end{pmatrix}.$$

Hint. Use the property that the matrices are symmetric and orthogonal. Moreover use the fact that the trace of an $n \times n$ matrix is equal to the sum of the eigenvalues of the matrix, i.e.

$$\operatorname{tr} X = \sum_{j=1}^{n} \lambda_j.$$

Solution. Since the matrices A and B are symmetric it follows that the eigenvalues λ_j are real. From the property that the matrices are orthogonal it follows that the eigenvalues are ± 1. From

$$\operatorname{tr} A = \lambda_1 + \lambda_2 + \lambda_3 = 1$$

we find that the eigenvalues of the matrix A are given by

$$\{1, 1, -1\}.$$

Since

$$\operatorname{tr} B = \lambda_1 + \lambda_2 + \lambda_3 + \lambda_4 = 0,$$

we find that the eigenvalues of B are given by

$$\{1, 1, -1, -1\}.$$

Problem 4.   (i) Let A be the symmetric matrix

$$A = \begin{pmatrix} 5 & -2 & -4 \\ -2 & 2 & 2 \\ -4 & 2 & 5 \end{pmatrix}.$$

Calculate the eigenvalues and eigenvectors of A. Are the eigenvectors orthogonal to each other? If not, try to find orthogonal eigenvectors using the *Gram–Schmidt algorithm*.

(ii) Let B be an arbitrary symmetric $n \times n$ matrix over **R**. Show that the eigenvectors which belong to different eigenvalues are orthogonal.

Solution (i) Since the matrix A is symmetric we find that the eigenvalues are real. The eigenvalues are determined by

$$\det(A - \lambda I) = 0. \tag{1}$$

From (1) we obtain the characteristic polynomial

$$-\lambda^3 + 12\lambda^2 - 21\lambda + 10 = 0.$$

The eigenvalues are

$$\lambda_1 = 1, \quad \lambda_2 = 1, \quad \lambda_3 = 10$$

with the corresponding eigenvectors

$$\mathbf{u}_1 = \begin{pmatrix} -1 \\ -2 \\ 0 \end{pmatrix}, \quad \mathbf{u}_2 = \begin{pmatrix} -1 \\ 0 \\ -1 \end{pmatrix}, \quad \mathbf{u}_3 = \begin{pmatrix} 2 \\ -1 \\ -2 \end{pmatrix}.$$

We find

$$(\mathbf{u}_1, \mathbf{u}_3) = 0, \quad (\mathbf{u}_2, \mathbf{u}_3) = 0, \quad (\mathbf{u}_1, \mathbf{u}_2) = 1$$

where $(,)$ denotes the *scalar product*, i.e.

$$(\mathbf{u}_j, \mathbf{u}_k) := \mathbf{u}_j{}^T \mathbf{u}_k$$

and ^{T} denotes the transpose. To apply the Gram–Schmidt algorithm we choose

$$\mathbf{u}'_1 = \mathbf{u}_1, \quad \mathbf{u}'_2 = \mathbf{u}_2 + \alpha \mathbf{u}_1$$

such that

$$\alpha = -\frac{(\mathbf{u}_1, \mathbf{u}_2)}{(\mathbf{u}_1, \mathbf{u}_1)} = -\frac{1}{5}.$$

Consequently,

$$\mathbf{u}'_2 = \begin{pmatrix} -\dfrac{4}{5} \\ \dfrac{2}{5} \\ -1 \end{pmatrix}.$$

The vectors $\mathbf{u}_1$, $\mathbf{u}'_2$, $\mathbf{u}_3$ are orthogonal.

(ii) From the eigenvalue equations

$$Bu_j = \lambda_j u_j, \quad Bu_k = \lambda_k u_k,$$

we obtain

$$u_k^T B u_j = \lambda_j u_k^T u_j, \quad u_j^T B u_k = \lambda_k u_j^T u_k.$$

Subtracting the two equations yields

$$0 = (\lambda_j - \lambda_k) u_k^T u_j \tag{2}$$

since $u_k^T u_j = u_j^T u_k$ and $u_k^T B u_j = u_j^T B u_k$. From (2) it follows that

$$u_k^T u_j \equiv (u_k, u_j) = 0$$

since $\lambda_j \neq \lambda_k$ by assumption.

Problem 5.   Let

$$A = \begin{pmatrix} -1 & 2 \\ 2 & -1 \end{pmatrix}.$$

Calculate $\exp(\epsilon A)$ where $\epsilon \in \mathbf{R}$.

Solution.   The matrix A is symmetric. Therefore, there exists an orthogonal matrix U such that UAU^{-1} is a diagonal matrix. The diagonal elements of UAU^{-1} are the eigenvalues of A. Since A is symmetric the eigenvalues are real. We set

$$D = UAU^{-1} \tag{1}$$

with

$$D = \begin{pmatrix} d_{11} & 0 \\ 0 & d_{22} \end{pmatrix}.$$

Then

$$\exp D = \begin{pmatrix} e^{d_{11}} & 0 \\ 0 & e^{d_{22}} \end{pmatrix}.$$

From (1) it follows that

$$e^{\epsilon D} = \exp(\epsilon U A U^{-1}) = U \exp(\epsilon A) U^{-1}.$$

Therefore

$$\exp(\epsilon A) = U^{-1}e^{\epsilon D}U.$$

The matrix U is constructed by means of the eigenvalues and normalized eigenvectors of A. The eigenvalues of A are given by

$$\lambda_1 = 1, \quad \lambda_2 = -3.$$

The corresponding normalized eigenvectors are

$$\frac{1}{\sqrt{2}}\begin{pmatrix} 1 \\ 1 \end{pmatrix}, \quad \frac{1}{\sqrt{2}}\begin{pmatrix} 1 \\ -1 \end{pmatrix}.$$

Consequently, the matrix U^* is given by

$$U^* = \frac{1}{\sqrt{2}}\begin{pmatrix} 1 & 1 \\ 1 & -1 \end{pmatrix}.$$

It follows that

$$U^* = U = U^{-1}.$$

Finally we arrive at

$$\exp(\epsilon A) = U^*e^{\epsilon D}U = \frac{1}{2}\begin{pmatrix} e^\epsilon + e^{-3\epsilon} & e^\epsilon - e^{-3\epsilon} \\ e^\epsilon - e^{-3\epsilon} & e^\epsilon + e^{-3\epsilon} \end{pmatrix}.$$

Remark. *The solution of the initial value problem of the autonomous system of linear ordinary differential equations*

$$\frac{du_1}{d\epsilon} = -u_1 + 2u_2,$$

$$\frac{du_2}{d\epsilon} = 2u_1 - u_2$$

is given by

$$\begin{pmatrix} u_1(\epsilon) \\ u_2(\epsilon) \end{pmatrix} = e^{\epsilon A}\begin{pmatrix} u_1(\epsilon = 0) \\ u_2(\epsilon = 0) \end{pmatrix}.$$

Problem 6.   Let

$$A = \begin{pmatrix} 0 & c & -b \\ -c & 0 & a \\ b & -a & 0 \end{pmatrix}$$

where a, b, $c \in \mathbf{R}$ and a, b, $c \neq 0$.

(i) Calculate the eigenvalues and eigenvectors of A.

(ii) Calculate $e^{\mu A}$ where $\mu \in \mathbf{R}$.

Solution.   (i) The matrix A is skew symmetric. Therefore the eigenvalues must be purely imaginary or zero. Since $\det A = 0$ and

$$\det A = \lambda_1 \lambda_2 \lambda_3$$

where λ_1, λ_2, λ_3 denote the eigenvalues we conclude that at least one of the eigenvalues must be zero. Thus the eigenvalues are 0, ki and $-ki$ where $k \in \mathbf{R}$. Eigenvalues of matrices over the field $\mathbf{R}$ occur in complex conjugate pairs. In the present case it also follows from the fact that

$$\mathrm{tr}A = \lambda_1 + \lambda_2 + \lambda_3 = 0.$$

To find k we have to solve

$$\det(A - \lambda I) = 0.$$

We obtain

$$\lambda^3 + \lambda(a^2 + b^2 + c^2) = 0.$$

Therefore

$$\lambda_1 = 0, \quad \lambda_2 = i\sqrt{a^2 + b^2 + c^2}, \quad \lambda_3 = -i\sqrt{a^2 + b^2 + c^2}.$$

The eigenvector of the eigenvalue $\lambda_1 = 0$ is determined by

$$A\begin{pmatrix} x_1 \\ x_2 \\ x_3 \end{pmatrix} = 0 \begin{pmatrix} x_1 \\ x_2 \\ x_3 \end{pmatrix} \equiv \begin{pmatrix} 0 \\ 0 \\ 0 \end{pmatrix}.$$

Consequently

$$cx_2 - bx_3 = 0, \quad -cx_1 + ax_3 = 0, \quad bx_1 - ax_2 = 0.$$

The solution is given by

$$
\begin{pmatrix} x_1 \\ x_2 \\ x_3 \end{pmatrix} = \begin{pmatrix} a \\ b \\ c \end{pmatrix} .
$$

The eigenvector of the eigenvalue $\lambda_2 = ik$ is determined by

$$
cx_2 - bx_3 = ikx_1 , \qquad -cx_1 + ax_3 = ikx_2 , \qquad bx_1 - ax_2 = ikx_3
$$

where

$$
k := \sqrt{a^2 + b^2 + c^2} .
$$

We find

$$
\begin{pmatrix} x_1 \\ x_2 \\ x_3 \end{pmatrix} = \begin{pmatrix} ac - ibk \\ bc + iak \\ c^2 - k^2 \end{pmatrix} .
$$

The eigenvector of the eigenvalue $\lambda_2 = -ik$ is determined by

$$
cx_2 - bx_3 = -ikx_1 , \qquad -cx_1 + ax_3 = -ikx_2 , \qquad bx_1 - ax_2 = -ikx_3 . \tag{1}
$$

Solving (1) yields

$$
\begin{pmatrix} x_1 \\ x_2 \\ x_3 \end{pmatrix} = \begin{pmatrix} ac + ibk \\ bc - iak \\ c^2 - k^2 \end{pmatrix} .
$$

ii) We calculate $e^{\mu A}$ in Chapter 12, Problem 2.

Problem 7. Let A be an $n \times n$ matrix. Show that

$$
\det(\exp A) \equiv \exp(\operatorname{tr} A) \tag{1}
$$

where $\operatorname{tr}(.)$ denotes the trace and det the determinant.

Solution. Any $n \times n$ matrix can be brought into a triangular form by a *similarity transformation*. This means there is an invertible $n \times n$ matrix R such that

$$R^{-1}AR = T \tag{2}$$

where T is a triangular matrix with diagonal elements t_{jj} which are the eigenvalues of A. We set $t_{jj} = \lambda_j$. From (2) it follows that

$$A = RTR^{-1}$$

and therefore

$$\exp A = \exp(RTR^{-1}) = R(\exp T)R^{-1}.$$

Since T is triangular, the diagonal elements of the k-th power of T are λ_j^k where k is a positive integer. Consequently, the diagonal elements of $\exp T$ are $\exp \lambda_j$. Since the determinant of a triangular matrix is equal to the product of its diagonal elements we find

$$\det(\exp T) = \exp(\lambda_1 + \lambda_2 + \cdots + \lambda_n) = \exp(\mathrm{tr}T).$$

Since

$$\mathrm{tr}T = \mathrm{tr}(R^{-1}AR) = \mathrm{tr}A$$

and

$$\det(\exp T) = \det(R(\exp T)R^{-1}) = \det(\exp(RTR^{-1})) = \det(\exp A),$$

we obtain (1).

Remark. *The identity*

$$\exp(RTR^{-1}) \equiv R(\exp T)R^{-1}$$

can easily be seen from

$$\exp(RTR^{-1}) = \sum_{k=0}^{\infty} \frac{(RTR^{-1})^k}{k!}$$

and $RR^{-1} = R^{-1}R = I$.

Problem 8.   Let A be a 4×4 symmetric matrix. Assume that the eigenvalues are given by 0, 1, 2, and 3 with the corresponding normalized eigenvectors

$$\frac{1}{\sqrt{2}}\begin{pmatrix} 1 \\ 0 \\ 0 \\ 1 \end{pmatrix},\quad \frac{1}{\sqrt{2}}\begin{pmatrix} 1 \\ 0 \\ 0 \\ -1 \end{pmatrix},\quad \frac{1}{\sqrt{2}}\begin{pmatrix} 0 \\ 1 \\ 1 \\ 0 \end{pmatrix},\quad \frac{1}{\sqrt{2}}\begin{pmatrix} 0 \\ 1 \\ -1 \\ 0 \end{pmatrix}.$$

Find the matrix A.

Solution.   Since A is a symmetric matrix there exists an orthogonal matrix U such that

$$D = UAU^T \tag{1}$$

where D is the diagonal matrix

$$D = \mathrm{diag}(0, 1, 2, 3)\,.$$

The matrix U^T is given by the normalized eigenvectors of A, i.e.

$$U^T = \frac{1}{\sqrt{2}}\begin{pmatrix} 1 & 1 & 0 & 0 \\ 0 & 0 & 1 & 1 \\ 0 & 0 & 1 & -1 \\ 1 & -1 & 0 & 0 \end{pmatrix}.$$

Thus

$$U = \frac{1}{\sqrt{2}}\begin{pmatrix} 1 & 0 & 0 & 1 \\ 1 & 0 & 0 & -1 \\ 0 & 1 & 1 & 0 \\ 0 & 1 & -1 & 0 \end{pmatrix}.$$

Since

$$U^T = U^{-1}\,,$$

we find from (1) that

$$A = U^T D U\,.$$

We obtain

$$A = \frac{1}{2} \begin{pmatrix} 1 & 0 & 0 & -1 \\ 0 & 5 & -1 & 0 \\ 0 & -1 & 5 & 0 \\ -1 & 0 & 0 & 1 \end{pmatrix}.$$

Problem 9.   Consider the infinite-dimensional symmetric matrix

$$A = \begin{pmatrix} 0 & 1 & 0 & 0 & & \cdots \\ 1 & 0 & 1 & 0 & & \cdots \\ 0 & 1 & 0 & 1 & & \cdots \\ & & \ddots & & \ddots & \\ & & & \ddots & & \ddots \\ & & & & \ddots & \end{pmatrix}.$$

In other words

$$A_{jk} = \begin{cases} 1 & \text{if } j = k+1, \\ 1 & \text{if } j = k-1, \\ 0 & \text{otherwise} \end{cases}$$

with j, $k = 1, 2, 3, \ldots$. Find the spectrum of this infinite-dimensional matrix.

Solution   Let A_n be the $n \times n$ truncated matrix of A. Then the eigenvalue problem for A_n is given by $A_n \mathbf{x} = \lambda \mathbf{x}$ with

$$A_n = \begin{pmatrix} 0 & 1 & 0 & 0 & \cdots & 0 & 0 & 0 \\ 1 & 0 & 1 & 0 & \cdots & 0 & 0 & 0 \\ 0 & 1 & 0 & 1 & \cdots & 0 & 0 & 0 \\ & & \ddots & & \ddots & & & \\ \vdots & & & \ddots & & \ddots & & \\ & & & & \ddots & & \ddots & \\ 0 & 0 & 0 & 0 & \cdots & 1 & 0 & 1 \\ 0 & 0 & 0 & 0 & \cdots & 0 & 1 & 0 \end{pmatrix}.$$

First we calculate the eigenvalues of A_n. Then we study A_n as $n \to \infty$. The eigenvalue problem leads to $D_n(\lambda) = 0$ where

$$D_n(\lambda) \equiv \det \begin{pmatrix} -\lambda & 1 & 0 & 0 & \cdots & 0 & 0 & 0 \\ 1 & -\lambda & 1 & 0 & \cdots & 0 & 0 & 0 \\ 0 & 1 & -\lambda & 1 & \cdots & 0 & 0 & 0 \\ & & \vdots & & & & & \\ 0 & 0 & 0 & 0 & \cdots & 1 & -\lambda & 1 \\ 0 & 0 & 0 & 0 & \cdots & 0 & 1 & -\lambda \end{pmatrix}.$$

We try to find a difference equation for $D_n(\lambda)$, where $n = 1, 2, \ldots$. We obtain

$$D_n(\lambda) = -\lambda \det \begin{pmatrix} -\lambda & 1 & 0 & 0 & \cdots & 0 & 0 & 0 \\ 1 & -\lambda & 1 & 0 & \cdots & 0 & 0 & 0 \\ 0 & 1 & -\lambda & 1 & \cdots & 0 & 0 & 0 \\ 0 & 0 & 1 & -\lambda & \cdots & 0 & 0 & 0 \\ & & \vdots & & & & & \\ 0 & 0 & 0 & 0 & \cdots & 1 & -\lambda & 1 \\ 0 & 0 & 0 & 0 & \cdots & 0 & 1 & -\lambda \end{pmatrix}$$

$$- \det \begin{pmatrix} 1 & 1 & 0 & 0 & \cdots & 0 & 0 & 0 \\ 0 & -\lambda & 1 & 0 & \cdots & 0 & 0 & 0 \\ 0 & 1 & -\lambda & 1 & \cdots & 0 & 0 & 0 \\ & \vdots & & & & & & \\ 0 & 0 & 0 & 0 & \cdots & 1 & -\lambda & 1 \\ 0 & 0 & 0 & 0 & \cdots & 0 & 1 & -\lambda \end{pmatrix}.$$

The first determinant on the right-hand side is equal to $D_{n-1}(\lambda)$. For the second determinant we find (expansion of the first row)

$$\det \begin{pmatrix} 1 & 1 & 0 & 0 & \cdots & 0 & 0 & 0 \\ 0 & -\lambda & 1 & 0 & \cdots & 0 & 0 & 0 \\ 0 & 1 & -\lambda & 1 & \cdots & 0 & 0 & 0 \\ \vdots & & & & & & & \\ 0 & 0 & 0 & 0 & \cdots & 1 & -\lambda & 1 \\ 0 & 0 & 0 & 0 & \cdots & 0 & 1 & -\lambda \end{pmatrix} = D_{n-2}(\lambda).$$

Consequently, we obtain a second order linear difference equation

$$D_n(\lambda) = -\lambda D_{n-1}(\lambda) - D_{n-2}(\lambda) \tag{1}$$

with the "initial condition"

$$D_1(\lambda) = -\lambda, \quad D_2(\lambda) = \lambda^2 - 1. \tag{2}$$

To solve this linear difference equation we make the ansatz

$$D_n(\lambda) = e^{in\theta} \tag{3}$$

where $n = 1, 2, \ldots$. Inserting the ansatz (3) into the difference Eq. (1) yields

$$e^{in\theta} = -\lambda e^{i(n-1)\theta} - e^{i(n-2)\theta}.$$

It follows that

$$e^{i\theta} = -\lambda - e^{-i\theta}.$$

Consequently

$$\lambda = -2\cos\theta.$$

The general solution to the difference Eq. (7) is given by

$$D_n(\lambda) = C_1 \cos(n\theta) + C_2 \sin(n\theta)$$

where C_1 and C_2 are constants and $\lambda = -2\cos\theta$. Imposing the initial conditions (2), it follows that

$$D_n(\lambda) = \frac{\sin(n+1)\theta}{\sin\theta}.$$

Since

$$D_n(\lambda) = 0,$$

we have to solve the equation

$$\frac{\sin(n+1)\theta}{\sin\theta} = 0.$$

The solutions to this equation are given by

$$\theta = \frac{k\pi}{n+1}$$

with $k = 1, 2, \ldots, n$. Since $\lambda = -2\cos\theta$, we find the eigenvalues

$$\lambda_k = -2\cos\left(\frac{k\pi}{n+1}\right)$$

with $k = 1, 2, \ldots, n$. Consequently, $|\lambda_k| \leq 2$. If $n \to \infty$, then there are infinitely many λ_k with $|\lambda_k| \leq 2$ and $\lambda_k - \lambda_{k+1} \to 0$. Therefore

$$\text{spectrum } A = [-2, 2],$$

i.e., we have a *continuous spectrum*.
Another approach to find the spectrum is as follows. Let

$$A = B + B^T$$

where

$$B = \begin{pmatrix} 0 & 0 & 0 & 0 & \cdots \\ 1 & 0 & 0 & 0 & \cdots \\ 0 & 1 & 0 & 0 & \cdots \\ 0 & 0 & 1 & 0 & \cdots \\ \vdots \end{pmatrix}, \quad B^T = \begin{pmatrix} 0 & 1 & 0 & 0 & \cdots \\ 0 & 0 & 1 & 0 & \cdots \\ 0 & 0 & 0 & 1 & \cdots \\ \vdots \end{pmatrix}.$$

Then

$$B^T B = I$$

where I is the infinite unit matrix. Notice that $BB^T \neq I$. We use the following notation

$$Cf = \lambda f \text{ means } \|Cf_n - \lambda f_n\| \to 0.$$

Now

$$Bf = \lambda f \Rightarrow B^T Bf = B^T \lambda f.$$

Therefore

$$\Rightarrow \text{ If} = \lambda B^T f \Rightarrow f = \lambda B^T f \Rightarrow B^T f = \frac{1}{\lambda} f.$$

From $Bf = \lambda f$ it also follows that

$$\|Bf\|^2 = (Bf, Bf) = (f, B^T B f) = (f, f) = \|f\|^2. \tag{4}$$

On the other hand

$$(Bf, Bf) = \bar{\lambda}\lambda(f, f) = |\lambda|^2 \|f\|^2. \tag{5}$$

Since $\|f\| > 0$ we find from (4) and (5) that

$$|\lambda|^2 = 1.$$

Therefore

$$Af = (B + B^T)f = (\lambda + \frac{1}{\lambda})f = (\lambda + \bar{\lambda})f = 2(\cos\phi)f.$$

This means

$$\bigwedge_{\phi \in \mathbf{R}} \|(A - 2(\cos\phi)I)f_n\| \to 0$$

or

$$A \begin{pmatrix} \sin\phi \\ \sin 2\phi \\ \sin 3\phi \\ \vdots \end{pmatrix} = 2\cos\phi \begin{pmatrix} \sin\phi \\ \sin 2\phi \\ \sin 3\phi \\ \vdots \end{pmatrix}. \tag{6}$$

The vector on the left-hand side of (6) is not an element of the Hilbert space $\ell_2(\mathbf{N})$. For the first two rows we have the identities

$$\sin 2\phi = 2\sin\phi\cos\phi$$

and

$$\sin\phi + \sin 3\phi = 2\cos\phi\sin 2\phi.$$

Here $\ell_2(\mathbf{N})$ is the *Hilbert space* of all infinite vectors (sequences)

$$\mathbf{u} = (u_1, u_2, \dots)^T$$

of complex numbers u_j such that

$$\sum_{j=1}^{\infty} |u_j|^2 < \infty .$$

Problem 10. Let $\hat{B}$ and $\hat{C}$ be two linear bounded operators with a discrete spectrum (for example two finite-dimensional matrices). Assume that

$$[\hat{B}, \hat{C}]_+ = 0 \qquad (1)$$

where

$$[\hat{B}, \hat{C}]_+ := \hat{B}\hat{C} + \hat{C}\hat{B} .$$

Let $\mathbf{u}$ be an eigenvector of both $\hat{B}$ and $\hat{C}$. What can be said about the corresponding eigenvalues?

Solution. From

$$\hat{B}\mathbf{u} = B\mathbf{u}$$

and

$$\hat{C}\mathbf{u} = C\mathbf{u}$$

where B and C are the eigenvalues of $\hat{B}$ and $\hat{C}$, respectively, we obtain using (1)

$$[\hat{B}, \hat{C}]_+\mathbf{u} = (\hat{B}\hat{C} + \hat{C}\hat{B})\mathbf{u} = \hat{B}(\hat{C}\mathbf{u}) + \hat{C}(\hat{B}\mathbf{u}) .$$

Thus

$$[\hat{B}, \hat{C}]_+\mathbf{u} = \hat{B}C\mathbf{u} + \hat{C}B\mathbf{u} = BC\mathbf{u} + CB\mathbf{u} .$$

Finally

$$[\hat{B}, \hat{C}]_+\mathbf{u} = 2BC\mathbf{u} = 0 .$$

Consequently, since $\mathbf{u} \neq \mathbf{0}$ we have the solutions

$$B = 0, \quad C \neq 0,$$
$$B \neq 0, \quad C = 0,$$
$$B = 0, \quad C = 0.$$

Remark. $[,]_+$ *is called the* anticommutator.

Problem 11.   Let A be an $n \times n$ matrix with eigenvalue λ.

(i) Show that λ^2 is an eigenvalue of A^2.
(ii) Show that e^λ is an eigenvalue of e^A.
(iii) Show that $\sin(\lambda)$ is an eigenvalue of $\sin(A)$.
(iv) Assume that A^{-1} exists. Show that $1/\lambda$ is an eigenvalue of A^{-1}.

Solution.   (i) From the eigenvalue equation

$$Ax = \lambda x \tag{1}$$

we obtain

$$A^2 x = A(Ax) = A(\lambda x) = \lambda A x = \lambda^2 x.$$

Obviously, $A^3 x = \lambda^3 x$ etc..

(ii) Using the expansion

$$e^A := \sum_{k=0}^{\infty} \frac{A^k}{k!}$$

and the result from (i) we find that e^A has the eigenvalue e^λ.

(iii) Using the expansion

$$\sin(A) := \sum_{k=0}^{\infty} \frac{(-1)^k A^{2k+1}}{(2k+1)!}$$

and the result from (i) we find that $\sin(\lambda)$ is an eigenvalue of $\sin(A)$.

(iv) From the eigenvalue Eq. (1) we find

$$A^{-1}(Ax) = A^{-1}(\lambda x).$$

Thus

$$x = \lambda A^{-1} x.$$

Since $x \neq 0$ we find that $1/\lambda$ is an eigenvalue of A^{-1}.

Problem 12.   A norm of an $n \times n$ matrix A over the real numbers can be defined as

$$\|A\| := \sup_{\|x\|=1} \|Ax\|$$

where $\| \ \|$ on the right-hand side denotes the Euclidean norm. Let

$$A = \begin{pmatrix} 1 & 1 \\ 2 & 2 \end{pmatrix}.$$

i) Find $\|A\|$.

ii) Find the eigenvalues of $A^T A$ and compare with the result of (i).

Solution.   (i) We apply the method of the Lagrange multiplier. Let

$$\mathbf{x} := \begin{pmatrix} x_1 \\ x_2 \end{pmatrix}, \quad \mathbf{x}^T = (x_1, x_2).$$

By matrix multiplication we obtain

$$\|A\mathbf{x}\|^2 = (A\mathbf{x})^T A\mathbf{x} = \mathbf{x}^T A^T A\mathbf{x} = 5x_1^2 + 10x_1x_2 + 5x_2^2.$$

The constraint is

$$\|\mathbf{x}\| = 1 \Leftrightarrow x_1^2 + x_2^2 = 1.$$

Thus we have

$$f(x_1, x_2) = 5x_1^2 + 10x_1x_2 + 5x_2^2 + \lambda(x_1^2 + x_2^2 - 1)$$

where λ is the Lagrange multiplier. From the equations

$$\frac{\partial f}{\partial x_1} = 10x_1 + 10x_2 + 2\lambda x_1 = 0,$$

$$\frac{\partial f}{\partial x_2} = 10x_1 + 10x_2 + 2\lambda x_2 = 0,$$

we find

$$x_1^2 = x_2^2 = \frac{1}{2}.$$

Thus the square of the norm is given by

$$\|A\|^2 = 5 \cdot \frac{1}{2} + 10 \cdot \frac{1}{2} + 5 \cdot \frac{1}{2} = 10.$$

ii) We have

$$A^T A = \begin{pmatrix} 5 & 5 \\ 5 & 5 \end{pmatrix}.$$

The rank of the matrix $A^T A$ is 1 and

$$\mathrm{tr}(A^T A) = 10.$$

Thus we find that the eigenvalues of $A^T A$ are given by 0 and 10. Thus the square of the norm is the largest eigenvalue of $A^T A$. Is this true in general?

Problem 13. Let

$$\mathbf{x} = \begin{pmatrix} x_1 \\ x_2 \\ x_3 \end{pmatrix}$$

where $x_j \in \mathbf{R}$. Show that at least one the eigenvalues of the 3×3 matrix

$$A := \begin{pmatrix} x_1 \\ x_2 \\ x_3 \end{pmatrix} (x_1 x_2 x_3)$$

is equal to zero.

Solution. We have

$$A := \begin{pmatrix} x_1 \\ x_2 \\ x_3 \end{pmatrix} (x_1 x_2 x_3) = \begin{pmatrix} x_1^2 & x_1 x_2 & x_1 x_3 \\ x_1 x_2 & x_2^2 & x_2 x_3 \\ x_1 x_3 & x_2 x_3 & x_3^2 \end{pmatrix}.$$

Straightforward calculation yields

$$\det(A) = 0.$$

Since

$$\det(A) = \lambda_1 \lambda_2 \lambda_3,$$

we can conclude that at least one eigenvalue is 0.

What are the other two eigenvalues?

Problem 14. Let A be an $n \times n$ matrix over $\mathbf{R}$. Assume that A^{-1} exists. To compute A^{-1} we can use the iteration

$$X_{t+1} = 2X_t - X_t A X_t, \quad t = 0, 1, 2, \dots$$

with $X_0 = \alpha A^T$. The scalar α is chosen such that $0 < \alpha < 2/\sigma_1$ with σ_1 to be the largest *singular value* of A. The singular values of a square matrix

B are the positive roots of the eigenvalues of the hermitian matrix B^*B (or B^TB if the matrix B is real). Apply the iteration to the matrix

$$A := \begin{pmatrix} 1 & 1 \\ 1 & 0 \end{pmatrix}.$$

Solution. The largest eigenvalue of A^TA is

$$\lambda = \frac{(3 + \sqrt{5})}{2}.$$

Thus we choose $\alpha = 1/2$ since

$$\frac{1}{2} < \frac{2\sqrt{2}}{\sqrt{3 + \sqrt{5}}}.$$

We obtain

$$X_1 = 2 \cdot \frac{1}{2} \begin{pmatrix} 1 & 1 \\ 1 & 0 \end{pmatrix} - \frac{1}{4} \begin{pmatrix} 1 & 1 \\ 1 & 0 \end{pmatrix} \begin{pmatrix} 1 & 1 \\ 1 & 0 \end{pmatrix} \begin{pmatrix} 1 & 1 \\ 1 & 0 \end{pmatrix} = \begin{pmatrix} \frac{1}{4} & \frac{1}{2} \\ \frac{1}{2} & -\frac{1}{4} \end{pmatrix}.$$

For X_2 we find

$$X_2 = \begin{pmatrix} \frac{3}{16} & \frac{11}{16} \\ \frac{11}{16} & -\frac{1}{2} \end{pmatrix}.$$

The series converges to the inverse of A

$$A^{-1} = \begin{pmatrix} 0 & 1 \\ 1 & -1 \end{pmatrix}.$$

Chapter 8

Transformations

Problem 1.   Let $f : [0,1] \mapsto [0,1]$ be defined by

$$f(x) = 4x(1-x) \tag{1}$$

and let $\phi : [0,1] \mapsto [0,1]$ be defined by

$$\phi(x) = \frac{2}{\pi} \arcsin \sqrt{x} .$$

Calculate

$$\tilde{f} = \phi \circ f \circ \phi^{-1}$$

where $\circ$ denotes the *composition of functions*.

Solution.   The inverse of ϕ is defined on $[0,1]$ and we find

$$\phi^{-1}(x) = \sin^2 \left(\frac{\pi x}{2} \right) \equiv \frac{1 - \cos(\pi x)}{2}$$

since

$$\arcsin \circ \sin y = y, \quad \sin \circ \arcsin x = x$$

where

$$y \in \left[-\frac{\pi}{2}, \frac{\pi}{2} \right] .$$

To find $\tilde{f}$, we set

$$y(x) = \sin^2 \left(\frac{\pi x}{2} \right) , \tag{2a}$$

$$v(y) = 4y(1-y), \tag{2b}$$

$$w(v) = \frac{2}{\pi} \arcsin \sqrt{v}. \tag{2c}$$

Inserting (2a) into (2b) gives

$$v(x) = 4\sin^2\left(\frac{\pi x}{2}\right)\left(1 - \sin^2\frac{\pi x}{2}\right)$$

$$= 4\sin^2\left(\frac{\pi x}{2}\right)\cos^2\left(\frac{\pi x}{2}\right) = \left(2\sin\left(\frac{\pi x}{2}\right)\cos\left(\frac{\pi x}{2}\right)\right)^2.$$

Now

$$\sqrt{v(x)} = 2\sin\left(\frac{\pi x}{2}\right)\cos\left(\frac{\pi x}{2}\right) = \sin \pi x.$$

For $x \in [0, \frac{1}{2}]$ we have $\pi x \in [0, \frac{\pi}{2}]$ and therefore we find for this range

$$w(x) = \frac{2}{\pi} \arcsin(\sin \pi x) = 2x.$$

For $x \in [\frac{1}{2}, 1]$ we have $\pi x \in [\frac{1}{2}\pi, \pi]$ and therefore we find for this range

$$w(x) = \frac{2}{\pi} \arcsin(\sin \pi x) = 2(1-x).$$

Consequently,

$$\tilde{f}(x) = \begin{cases} 2x & \text{if } x \in \left[0, \dfrac{1}{2}\right] \\[2mm] 2(1-x) & \text{if } x \in \left[\dfrac{1}{2}, 1\right]. \end{cases} \tag{3}$$

Remark. *The mapping (1) is called the* logistic map *and mapping (3) is called the* tent map. *Both play an important role in the study of nonlinear systems with chaotic behaviour.*

Problem 2. Show that the one-dimensional wave equation

$$\frac{1}{c^2}\frac{\partial^2 u}{\partial t^2} = \frac{\partial^2 u}{\partial x^2} \tag{1}$$

is invariant under the transformation

$$\begin{pmatrix} x'(x,t) \\ ct'(x,t) \end{pmatrix} = \begin{pmatrix} \cosh \epsilon & \sinh \epsilon \\ \sinh \epsilon & \cosh \epsilon \end{pmatrix}\begin{pmatrix} x \\ ct \end{pmatrix}, \tag{2a}$$

$$u'(x'(x,t), t'(x,t)) = u(x,t) \tag{2b}$$

where ϵ is a real parameter. Equation (2a) is called the *Lorentz transformation*.

Hint. We have to show that

$$\frac{1}{c^2}\frac{\partial^2 u'}{\partial t'^2} = \frac{\partial^2 u'}{\partial x'^2}. \tag{3}$$

Solution.   Applying the *chain rule* gives

$$\frac{\partial u'}{\partial t} = \frac{\partial u'}{\partial x'}\frac{\partial x'}{\partial t} + \frac{\partial u'}{\partial t'}\frac{\partial t'}{\partial t} = \frac{\partial u}{\partial t}$$

and

$$\frac{\partial u'}{\partial x} = \frac{\partial u'}{\partial x'}\frac{\partial x'}{\partial x} + \frac{\partial u'}{\partial t'}\frac{\partial t'}{\partial x} = \frac{\partial u}{\partial x}.$$

It follows that

$$\frac{\partial^2 u'}{\partial t^2} = \left(\frac{\partial^2 u'}{\partial x'^2}\frac{\partial x'}{\partial t} + \frac{\partial^2 u'}{\partial x'\partial t'}\frac{\partial t'}{\partial t}\right)\frac{\partial x'}{\partial t} + \left(\frac{\partial^2 u'}{\partial x'\partial t'}\frac{\partial x'}{\partial t} + \frac{\partial^2 u'}{\partial t'^2}\frac{\partial t'}{\partial t}\right)\frac{\partial t'}{\partial t} = \frac{\partial^2 u}{\partial t^2}$$

and

$$\frac{\partial^2 u'}{\partial x^2} = \left(\frac{\partial^2 u'}{\partial x'^2}\frac{\partial x'}{\partial x} + \frac{\partial^2 u'}{\partial x'\partial t'}\frac{\partial t'}{\partial x}\right)\frac{\partial x'}{\partial x} + \left(\frac{\partial^2 u'}{\partial x'\partial t'}\frac{\partial x'}{\partial x} + \frac{\partial^2 u'}{\partial t'^2}\frac{\partial t'}{\partial x}\right)\frac{\partial t'}{\partial x}$$

$$= \frac{\partial^2 u}{\partial x^2}.$$

Since

$$\frac{\partial x'}{\partial t} = c\sinh\epsilon, \qquad \frac{\partial x'}{\partial x} = \cosh\epsilon$$

and

$$\frac{\partial t'}{\partial t} = \cosh\epsilon, \qquad \frac{\partial t'}{\partial x} = \frac{1}{c}\sinh\epsilon,$$

we obtain

$$c^2\frac{\partial^2 u'}{\partial x'^2}\sinh^2\epsilon + 2c\frac{\partial^2 u'}{\partial x'\partial t'}\sinh\epsilon\cosh\epsilon + \frac{\partial^2 u'}{\partial t'^2}\cosh^2\epsilon = \frac{\partial^2 u}{\partial t^2} \tag{4}$$

and

$$\frac{\partial^2 u'}{\partial x'^2}\cosh^2\epsilon + \frac{2}{c}\frac{\partial^2 u'}{\partial x'\partial t'}\sinh\epsilon\cosh\epsilon + \frac{1}{c^2}\frac{\partial^2 u'}{\partial t'^2}\sinh^2\epsilon = \frac{\partial^2 u}{\partial x^2}. \tag{5}$$

nserting (4) and (5) into the wave Eq. (1) we find (3), where we have used he identity

$$\cosh^2 \epsilon - \sinh^2 \epsilon \equiv 1.$$

Thus (1) is *invariant* under the transformation (2).

Problem 3.   Let

$$\frac{d^2\phi}{dt^2} = 0.$$

(1)

i) Let

$$u(t) = \frac{d\ln\phi}{dt}.$$

(2a)

Derive the differential equation for u.

ii) Let

$$u(t) = \frac{d^2\ln\phi}{dt^2}.$$

(2b)

Derive the differential equation for u.

iii) Let

$$u(t) = \frac{d^3\ln\phi}{dt^3}.$$

(2c)

Derive the differential equation for u.

Hint. Use the following notation:

$$A_1(t) := \int^t u(t_0)dt_0,$$

(3a)

$$A_2(t) := \int^t dt_1 \int^{t_1} u(t_0)dt_0,$$

(3b)

$$A_3(t) := \int^t dt_2 \int^{t_2} dt_1 \int^{t_1} u(t_0)dt_0.$$

(3c)

Solution.   (i) From (3a) through (3c) we find

$$\frac{dA_3}{dt} = A_2, \quad \frac{dA_2}{dt} = A_1, \quad \frac{dA_1}{dt} = A_0 \equiv u$$

and

$$\frac{de^{A_n}}{dt} = A_{n-1} e^{A_n}$$

where $n = 1, 2, 3$. From (2a) we obtain

$$u(t) = \frac{1}{\phi} \frac{d\phi}{dt}.$$

Therefore

$$\phi(t) = e^{A_1}.$$

Taking the derivative we find

$$\frac{d\phi}{dt} = e^{A_1} u$$

and

$$\frac{d^2\phi}{dt^2} = e^{A_1} u^2 + e^{A_1} \frac{du}{dt} \equiv e^{A_1} \left(\frac{du}{dt} + u^2 \right).$$

Consequently (2a) leads to the nonlinear differential equation

$$\frac{du}{dt} + u^2 = 0.$$

This is a *Riccati equation*.

(ii) From (2b) we obtain

$$\phi(t) = e^{A_2}.$$

The derivatives of ϕ give

$$\frac{d\phi}{dt} = e^{A_2} A_1$$

and

$$\frac{d^2\phi}{dt^2} = e^{A_2}(A_1^2 + u).$$

Therefore

$$A_1^2 + u = 0. \tag{4}$$

Taking the derivative of (4) leads to

$$2A_1 u + \frac{du}{dt} = 0 \tag{5}$$

and

$$2u^2 + 2A_1\frac{du}{dt} + \frac{d^2u}{dt^2} = 0. \tag{6}$$

To eliminate A_1 we multiply (6) with u and insert (5). This leads to the nonlinear differential equation

$$u\frac{d^2u}{dt^2} - \left(\frac{du}{dt}\right)^2 + 2u^3 = 0.$$

iii) From (2c) we obtain

$$\phi(t) = e^{A_3}.$$

Consequently

$$\frac{d^2\phi}{dt^2} = e^{A_3}(A_2^2 + A_1).$$

It follows that

$$A_1 + A_2^2 = 0 \tag{7}$$

and

$$u + 2A_1A_2 = 0. \tag{8}$$

From (7) and (8) we obtain

$$uA_2 = 2A_1^2. \tag{9}$$

From (8) it follows that

$$\frac{du}{dt} + 2uA_2 + 2A_1^2 = 0. \tag{10}$$

Inserting (9) into (10) gives

$$\frac{du}{dt} + 6A_1^2 = 0.$$

Therefore

$$\frac{d^2u}{dt^2} + 12A_1u = 0 \tag{11}$$

and

$$\frac{d^3u}{dt^3} + 12A_1\frac{du}{dt} + 12u^2 = 0. \tag{12}$$

Multiplying (12) by u and inserting A_1 from (11) leads to the nonlinear differential equation

$$u\frac{d^3u}{dt^3} - \frac{du}{dt}\frac{d^2u}{dt^2} + 12u^3 = 0.$$

Remark.   *The technique to find nonlinear differential equations from a linear differential equation using* (1) *and* (3) *can be extended to*

$$A_n(t) := \int^t dt_{n-1} \int^{t_{n-1}} dt_{n-2} \cdots \int^{t_1} u(t_0)dt_0.$$

Problem 4.   Let

$$\frac{d^2w}{dz^2} + p_1(z)\frac{dw}{dz} + p_2(z)w = 0$$

be a second-order linear differential equation in the complex domain. Let

$$\tilde{z}(z) = z,$$

$$v(\tilde{z}(z)) = w(z)\exp\left(\frac{1}{2}\int^z p_1(s)ds\right).$$

Find the differential equation for $v(\tilde{z})$.

Solution.   Let

$$R(z) := \frac{1}{2}\int^z p_1(s)ds.$$

Then

$$\frac{dv}{dz} = \frac{dv}{d\tilde{z}}\frac{d\tilde{z}}{dz} = \frac{dv}{d\tilde{z}} = \frac{dw}{dz}e^R + \frac{1}{2}p_1we^R \tag{1}$$

where we have used that $dR/dz = p_1/2$. Since $d\tilde{z}/dz = 1$ we obtain from (1)

$$\frac{d^2v}{d\tilde{z}^2} = \frac{d^2w}{dz^2}e^R + p_1\frac{dw}{dz}e^R + \frac{1}{2}\frac{dp_1}{dz}we^R + \frac{1}{4}p_1^2we^R.$$

Consequently

$$\frac{d^2v}{d\tilde{z}^2} + I(\tilde{z})v = 0$$

where

$$I(\tilde{z}) = p_2(\tilde{z}) - \frac{1}{2}\frac{dp_1(\tilde{z})}{d\tilde{z}} - \frac{1}{4}p_1^2(\tilde{z}).$$

Problem 5.   Consider the nonlinear partial differential equation

$$\left(\frac{\partial u}{\partial t}\right)^2 \frac{\partial^2 u}{\partial x^2} + \left(\frac{\partial u}{\partial x}\right)^2 \frac{\partial^2 u}{\partial t^2} - 2\frac{\partial u}{\partial x}\frac{\partial u}{\partial t}\frac{\partial^2 u}{\partial x \partial t} = 0. \tag{1}$$

i) Show that the partial differential Eq. (1) can be linearized with the following transformation

$$\epsilon(x, t) = \frac{\partial u}{\partial x}, \tag{2a}$$

$$\eta(x, t) = \frac{\partial u}{\partial t}, \tag{2b}$$

$$x(\epsilon, \eta) = \frac{\partial W}{\partial \epsilon}, \tag{2c}$$

$$t(\epsilon, \eta) = \frac{\partial W}{\partial \eta}, \tag{2d}$$

$$u(x, t) + W(\epsilon(x, t), \eta(x, t)) = x\epsilon(x, t) + t\eta(x, t). \tag{2e}$$

Remark.   *Transformation* (2) *is called the* Legendre *transformation*.

ii) Give the solution to the linearized equation.

Solution.   (i) Since

$$\epsilon = \frac{\partial u}{\partial x}(x(\epsilon, \eta), t(\epsilon, \eta)),$$

we obtain by taking the derivative with respect to ϵ

$$1 = \frac{\partial^2 u}{\partial x^2}\frac{\partial x}{\partial \epsilon} + \frac{\partial^2 u}{\partial x \partial t}\frac{\partial t}{\partial \epsilon} = \frac{\partial^2 u}{\partial x^2}\frac{\partial^2 W}{\partial \epsilon^2} + \frac{\partial^2 u}{\partial x \partial t}\frac{\partial^2 W}{\partial \epsilon \partial \eta}$$

where we have used (2c) and (2d). Analogously, we find

$$0 = \frac{\partial^2 u}{\partial x \partial t}\frac{\partial^2 W}{\partial \epsilon^2} + \frac{\partial^2 u}{\partial t^2}\frac{\partial^2 W}{\partial \epsilon \partial \eta},$$

$$0 = \frac{\partial^2 u}{\partial x^2}\frac{\partial^2 W}{\partial \epsilon \partial \eta} + \frac{\partial^2 u}{\partial x \partial t}\frac{\partial^2 W}{\partial \eta^2},$$

$$1 = \frac{\partial^2 u}{\partial x \partial t}\frac{\partial^2 W}{\partial \epsilon \partial \eta} + \frac{\partial^2 u}{\partial t^2}\frac{\partial^2 W}{\partial \eta^2}.$$

If we set

$$\rho := \frac{\partial^2 u}{\partial x^2}\frac{\partial^2 u}{\partial t^2} - \left(\frac{\partial^2 u}{\partial x \partial t}\right)^2 ,$$

we obtain

$$\frac{\partial^2 W}{\partial \epsilon^2}\frac{\partial^2 W}{\partial \eta^2} - \left(\frac{\partial^2 W}{\partial \epsilon \partial \eta}\right)^2 = \frac{1}{\rho}.$$

Inserting these expressions into (1) we obtain the linear equation

$$\epsilon^2\frac{\partial^2 W}{\partial \epsilon^2} + 2\epsilon\eta\frac{\partial^2 W}{\partial \epsilon \partial \eta} + \eta^2\frac{\partial^2 W}{\partial \eta^2} = 0. \tag{3}$$

(ii) Since (3) can be written in the form

$$\left(\left(\epsilon\frac{\partial}{\partial \epsilon} + \eta\frac{\partial}{\partial \eta}\right)^2 - \left(\epsilon\frac{\partial}{\partial \epsilon} + \eta\frac{\partial}{\partial \eta}\right)\right) W = 0,$$

we find that the general solution is given by

$$W(\epsilon, \eta) = G\left(\frac{\eta}{\epsilon}\right) + \eta H\left(\frac{\epsilon}{\eta}\right)$$

where G and H are two arbitrary smooth functions.

Problem 6.   The nonlinear ordinary differential equation

$$\frac{d^2 u}{dt^2} + 3\frac{du}{dt}u + u^3 = 0 \tag{1}$$

occurs in the investigation of univalued functions defined by second-order differential equations and in the study of the *modified Emden equation*.

(i) Show that (1) can be linearized with the help of the transformations

$$U(T) = u^2(t(T)), \quad dT = u(t(T))dt(T), \tag{2}$$

$$u(t) = \frac{1}{\phi(t)}\frac{d\phi(t)}{dt}. \tag{3}$$

(ii) Find the general solution to (1).

Solution.   (i) First we consider the transformation given by (2). Since

$$\frac{dU}{dT} = 2u\frac{du}{dt}\frac{dt}{dT} = 2\frac{du}{dt}$$

and

$$\frac{d^2U}{dT^2} = 2\frac{d^2u}{dt^2}\frac{dt}{dT} = \frac{2}{u}\frac{d^2u}{dt^2},$$

we obtain

$$\frac{u}{2}\frac{d^2U}{dT^2} + \frac{3u}{2}\frac{dU}{dT} + uU = 0.$$

With $u \neq 0$ it follows that

$$\frac{d^2U}{dT^2} + 3\frac{dU}{dT} + 2U = 0.$$

Now we consider the transformation given by (3). Since

$$\frac{du}{dt} = \frac{1}{\phi^2}\left(\frac{d^2\phi}{dt^2}\phi - \left(\frac{d\phi}{dt}\right)^2\right)$$

and

$$\frac{d^2u}{dt^2} = \frac{1}{\phi^3}\left(\frac{d^3\phi}{dt^3}\phi^2 - 3\phi\frac{d^2\phi}{dt^2}\frac{d\phi}{dt} + 2\left(\frac{d\phi}{dt}\right)^3\right),$$

we obtain (with $\phi \neq 0$)

$$\frac{d^3\phi}{dt^3} = 0. \tag{4}$$

i) The general solution to (1) can now easily be found with the help of (4) and transformation (2). The general solution to (4) is given by

$$\phi(t) = C_1t^2 + C_2t + C_3 \tag{5}$$

where C_1, C_2 and C_3 are the constants of integration. Inserting (5) into (3) yields the general solution

$$u(t) = \frac{2t + (C_2/C_1)}{t^2 + (C_2/C_1)t + (C_3/C_1)}$$

where $C_1 \neq 0$. If $C_1 = 0$ and $C_2 \neq 0$, then

$$u(t) = \frac{1}{t + C_3/C_2}.$$

$C_1 = C_2 = 0$ and $C_3 \neq 0$, then $u(t) = 0$.

Problem 7.   Consider the ordinary differential equation

$$\frac{d^2u}{dt^2} + \lambda(\phi(t))^{2m-2}u = 0 \tag{1}$$

where $m = 1, 2, \ldots,$ and λ is a real parameter The smooth function ϕ satisfies

$$\frac{d^2\phi}{dt^2} + (\phi(t))^{2m-1} = 0 \tag{2}$$

with

$$\phi(0) = 1, \quad \frac{d\phi(0)}{dt} = 0. \tag{3}$$

Perform the transformation

$$z(t) = (\phi(t))^{2m}, \tag{4a}$$

$$\bar{u}(z(t)) = u(t). \tag{4b}$$

Solution.   First we notice that the integration of (2) yields

$$\frac{1}{2}\left(\frac{d\phi}{dt}\right)^2 + \frac{1}{2m}\phi^{2m} = \frac{1}{2m} \tag{5}$$

where we have taken into account the initial conditions (3). From (4b) it follows that

$$\frac{d\bar{u}}{dt} = \frac{d\bar{u}}{dz}\frac{dz}{dt} = \frac{du}{dt}$$

and

$$\frac{d^2\bar{u}}{dt^2} = \frac{d^2\bar{u}}{dz^2}\left(\frac{dz}{dt}\right)^2 + \frac{d\bar{u}}{dz}\frac{d^2z}{dt^2} = \frac{d^2u}{dt^2}. \tag{6}$$

From (4b) we find that

$$\frac{dz}{dt} = 2m(\phi(t))^{2m-1}\frac{d\phi}{dt}. \tag{7}$$

Consequently,

$$\left(\frac{dz}{dt}\right)^2 = 4m^2(\phi(t))^{4m-2}\left(\frac{d\phi}{dt}\right)^2 = 4m(\phi(t))^{2m-2}z(1-z) \tag{8}$$

where we have used (5). From (7) we also obtain

$$\frac{d^2 z}{dt^2} = 2(\phi(t))^{2m-2}((1 - 3m)z + 2m - 1).\tag{9}$$

Again we have used (5). Inserting (7) through (9) into (6) yields

$$z(1 - z)\frac{d^2\bar{u}}{dz^2} + (\gamma - (\alpha + \beta + 1)z)\frac{d\bar{u}}{dz} - \alpha\beta\bar{u} = 0\tag{10}$$

where

$$\alpha + \beta = \frac{m - 1}{2m}, \quad \alpha\beta = \frac{-\lambda}{4m}, \quad \gamma = \frac{2m - 1}{2m}.$$

Equation (10) has three singular points at

$$z = 0, 1 \text{ and } \infty.$$

Remark. *Equation* (10) *is called the* hypergeometric equation.

Problem 8. Let

$$\bar{t}(x, t) = t,\tag{1a}$$

$$\bar{x}(x, t) = \int^x \frac{1}{u(s, t)}ds,\tag{1b}$$

$$\bar{u}(\bar{x}(x, t), \bar{t}(x, t)) = u(x, t).\tag{1c}$$

Let

$$\frac{\partial u}{\partial t} = u^2 \frac{\partial^2 u}{\partial x^2}.\tag{2}$$

Find the partial differential equation for $\bar{u}(\bar{x}, \bar{t})$.

Solution. From (1a) and (1b) we obtain

$$\frac{\partial \bar{t}}{\partial t} = 1,$$

$$\frac{\partial \bar{t}}{\partial x} = 0,$$

$$\frac{\partial \bar{x}}{\partial x} = \frac{1}{u(x, t)},$$

$$\frac{\partial \bar{x}}{\partial t} = -\int^x \frac{u_t(s, t)}{u^2(s, t)}ds = -\int^x \frac{\partial^2 u(s, t)}{\partial s^2}ds = -\frac{\partial u}{\partial x},$$

where $u_t \equiv \partial u/\partial t$. From (1c) we find

$$\frac{\partial \bar{u}}{\partial x} = \frac{\partial \bar{u}}{\partial \bar{x}}\frac{\partial \bar{x}}{\partial x} + \frac{\partial \bar{u}}{\partial \bar{t}}\frac{\partial \bar{t}}{\partial x} = \frac{\partial u}{\partial x}.$$

Therefore

$$\frac{\partial \bar{u}}{\partial \bar{x}} = u\frac{\partial u}{\partial x}. \tag{3}$$

Analogously,

$$\frac{\partial \bar{u}}{\partial t} = \frac{\partial \bar{u}}{\partial \bar{x}}\frac{\partial \bar{x}}{\partial t} + \frac{\partial \bar{u}}{\partial \bar{t}}\frac{\partial \bar{t}}{\partial t} = \frac{\partial u}{\partial t}$$

and therefore

$$-\frac{\partial \bar{u}}{\partial \bar{x}}\frac{\partial u}{\partial x} + \frac{\partial \bar{u}}{\partial \bar{t}} = \frac{\partial u}{\partial t}. \tag{4}$$

From (3) we obtain

$$\frac{\partial}{\partial x}\left(\frac{\partial \bar{u}}{\partial \bar{x}}\right) = \left(\frac{\partial u}{\partial x}\right)^2 + u\frac{\partial^2 u}{\partial x^2}.$$

It follows that

$$\left(\frac{\partial^2 \bar{u}}{\partial \bar{x}^2}\frac{\partial \bar{x}}{\partial x} + \frac{\partial^2 \bar{u}}{\partial \bar{x}\partial \bar{t}}\frac{\partial \bar{t}}{\partial x}\right) = \left(\frac{\partial u}{\partial x}\right)^2 + u\frac{\partial^2 u}{\partial x^2}.$$

Since $\partial \bar{t}/\partial x = 0$ and

$$\frac{\partial \bar{x}}{\partial x} = u^{-1},$$

we obtain

$$\frac{\partial^2 \bar{u}}{\partial \bar{x}^2} = u\left(\frac{\partial u}{\partial x}\right)^2 + u^2\frac{\partial^2 u}{\partial x^2}. \tag{5}$$

Inserting (1c), (3), (4) and (5) into (2) gives

$$\frac{\partial \bar{u}}{\partial \bar{t}} = \frac{\partial^2 \bar{u}}{\partial \bar{x}^2}. \tag{6}$$

Equation (6) is the *linear diffusion equation*.

Problem 9.   The nonlinear partial differential equation

$$\frac{\partial^2 u}{\partial t^2} - \frac{\partial^2 u}{\partial x^2} + f_1(x+t)f_2(x-t)\sin u = 0$$

s an extended one-dimensional *sine-Gordon equation*, where f_1 and f_2 are wo smooth functions. Let

$$\xi(x,t) := \int^{x+t} f_1(s)ds\,, \tag{1a}$$

$$\eta(x,t) := \int^{x-t} f_2(s)ds\,, \tag{1b}$$

$$\bar{u}(\xi(x,t),\eta(x,t)) := u(x,t)\,. \tag{1c}$$

Find the partial differential equation for $\bar{u}(\xi,\eta)$.

Solution. From (1a) and (1b) we obtain

$$\frac{\partial\xi}{\partial x} = f_1(x+t)\,, \quad \frac{\partial\eta}{\partial x} = f_2(x-t)\,, \tag{2a}$$

$$\frac{\partial\xi}{\partial t} = f_1(x+t)\,, \quad \frac{\partial\eta}{\partial t} = -f_2(x-t)\,. \tag{2b}$$

Since

$$\frac{\partial\bar{u}}{\partial x} = \frac{\partial\bar{u}}{\partial\xi}\frac{\partial\xi}{\partial x} + \frac{\partial\bar{u}}{\partial\eta}\frac{\partial\eta}{\partial x} = \frac{\partial u}{\partial x}\,,$$

we obtain using (2a)

$$\frac{\partial\bar{u}}{\partial\xi}f_1(x+t) + \frac{\partial\bar{u}}{\partial\eta}f_2(x-t) = \frac{\partial u}{\partial x}\,. \tag{3}$$

Analogously,

$$\frac{\partial\bar{u}}{\partial t} = \frac{\partial\bar{u}}{\partial\xi}\frac{\partial\xi}{\partial t} + \frac{\partial\bar{u}}{\partial\eta}\frac{\partial\eta}{\partial t} = \frac{\partial u}{\partial t}$$

and

$$\frac{\partial\bar{u}}{\partial\xi}f_1(x+t) - \frac{\partial\bar{u}}{\partial\eta}f_2(x-t) = \frac{\partial u}{\partial t}\,. \tag{4}$$

From (3) we find

$$\frac{\partial}{\partial x}\left(\frac{\partial\bar{u}}{\partial\xi}f_1(x+t) + \frac{\partial\bar{u}}{\partial\eta}f_2(x-t)\right) = \frac{\partial^2 u}{\partial x^2}\,.$$

Therefore

$$\frac{\partial^2\bar{u}}{\partial\xi^2}f_1^2 + 2\frac{\partial^2\bar{u}}{\partial\xi\partial\eta}f_1 f_2 + \frac{\partial^2\bar{u}}{\partial\eta^2}f_2^2 + \frac{\partial\bar{u}}{\partial\xi}f_1' + \frac{\partial\bar{u}}{\partial\eta}f_2' = \frac{\partial^2 u}{\partial x^2}$$

where $f_1' \equiv df_1(s)/ds$. Analogously, from (4) we obtain

$$\frac{\partial}{\partial t}\left(\frac{\partial \bar{u}}{\partial \xi}f_1(x+t) - \frac{\partial \bar{u}}{\partial \eta}f_2(x-t)\right) = \frac{\partial^2 u}{\partial t^2}.$$

Therefore

$$\frac{\partial^2 \bar{u}}{\partial \xi^2}f_1^2 - 2\frac{\partial^2 \bar{u}}{\partial \xi \partial \eta}f_1 f_2 + \frac{\partial^2 \bar{u}}{\partial \eta^2}f_2^2 + \frac{\partial \bar{u}}{\partial \xi}f_1' + \frac{\partial \bar{u}}{\partial \eta}f_2' = \frac{\partial^2 u}{\partial t^2}.$$

Consequently, we obtain

$$4\frac{\partial^2 \bar{u}}{\partial \eta \partial \xi} - \sin\bar{u} = 0.$$

Problem 10.   Let

$$\bar{t}(x,t) = t, \tag{1a}$$

$$\bar{x}(x,t) = u(x,t), \tag{1b}$$

$$\bar{u}(\bar{x}(x,t),\bar{t}(x,t)) = x, \tag{1c}$$

and

$$\frac{\partial u}{\partial t} = \frac{\partial^2 u}{\partial x^2} + u\frac{\partial u}{\partial x}. \tag{2}$$

Find the partial differential equation for $\bar{u}(\bar{x},\bar{t})$.

Solution.   From (1a) and (1b) we obtain

$$\frac{\partial \bar{t}}{\partial t} = 1, \quad \frac{\partial \bar{t}}{\partial x} = 0, \quad \frac{\partial \bar{x}}{\partial t} = \frac{\partial u}{\partial t}, \quad \frac{\partial \bar{x}}{\partial x} = \frac{\partial u}{\partial x}.$$

From (1c) we find

$$\frac{\partial \bar{u}}{\partial x} = \frac{\partial \bar{u}}{\partial \bar{x}}\frac{\partial \bar{x}}{\partial x} + \frac{\partial \bar{u}}{\partial \bar{t}}\frac{\partial \bar{t}}{\partial x} = 1.$$

Therefore

$$\frac{\partial \bar{u}}{\partial \bar{x}}\frac{\partial u}{\partial x} = 1. \tag{3}$$

Analogously,

$$\frac{\partial \bar{u}}{\partial t} = \frac{\partial \bar{u}}{\partial \bar{x}}\frac{\partial \bar{x}}{\partial t} + \frac{\partial \bar{u}}{\partial \bar{t}}\frac{\partial \bar{t}}{\partial t} = 0$$

and therefore

$$\frac{\partial \bar{u}}{\partial \bar{x}}\frac{\partial u}{\partial t} + \frac{\partial \bar{u}}{\partial \bar{t}} = 0.$$

From (3) we obtain

$$\frac{\partial}{\partial x}\left(\frac{\partial \bar{u}}{\partial \bar{x}}\frac{\partial u}{\partial x}\right) = 0.$$

It follows that

$$\left(\frac{\partial^2 \bar{u}}{\partial \bar{x}^2}\frac{\partial \bar{x}}{\partial x} + \frac{\partial^2 \bar{u}}{\partial \bar{x}\partial \bar{t}}\frac{\partial \bar{t}}{\partial x}\right)\frac{\partial u}{\partial x} + \frac{\partial \bar{u}}{\partial \bar{x}}\frac{\partial^2 u}{\partial x^2} = 0.$$

Consequently

$$\frac{\partial^2 \bar{u}}{\partial \bar{x}^2}\left(\frac{\partial u}{\partial x}\right)^2 + \frac{\partial \bar{u}}{\partial \bar{x}}\frac{\partial^2 u}{\partial x^2} = 0.$$

To summarize

$$u = \bar{x},\tag{4a}$$

$$\frac{\partial u}{\partial t} = -\frac{\bar{u}_{\bar{t}}}{\bar{u}_{\bar{x}}},\tag{4b}$$

$$\frac{\partial u}{\partial x} = \frac{1}{\bar{u}_{\bar{x}}},\tag{4c}$$

$$\frac{\partial^2 u}{\partial x^2} = -\frac{\bar{u}_{\bar{x}\bar{x}}}{\bar{u}_{\bar{x}}^3},\tag{4d}$$

where $\bar{u}_{\bar{t}} = \partial \bar{u}/\partial \bar{t}$ etc.. Inserting (4a) through (4d) into (2) gives

$$\left(\frac{\partial \bar{u}}{\partial \bar{x}}\right)^2 \frac{\partial \bar{u}}{\partial \bar{t}} = \frac{\partial^2 \bar{u}}{\partial \bar{x}^2} - \left(\frac{\partial \bar{u}}{\partial \bar{x}}\right)^2 \bar{x}.$$

Problem 11.   (i) Consider the potential

$$U(q) = \frac{1}{4}q^4 + \frac{\mu_0}{3}q^3 + \frac{\mu_1}{2}q^2 + \mu_2 q$$

where μ_0, μ_1 and μ_2 are real parameters. Show that the term $\mu_0 q^3/3$ can be eliminated with the help of the transformation

$$\tilde{q}(q) = q + \delta.\tag{1}$$

(ii) Let

$$U(q, \mu_1, \mu_2) = \frac{1}{4}q^4 + \frac{\mu_1}{2}q^2 + \mu_2 q. \qquad (2)$$

Find the parameter region where the potential $U(q) = q^4/4$ changes qualitatively with respect to μ_1 and μ_2.

(iii) Show that the *van der Waals equation*

$$(V - b)\left(P + \frac{a}{V^2}\right) = RT$$

can be expressed as

$$q^3 + \mu_1 q + \mu_2 = 0$$

where a, b and R are positive constants. Here P, V and T are the pressure, volume and temperature, respectively.

Solution.   (i) From (1) we obtain

$$q^3 = \tilde{q}^3 - 3\delta\tilde{q}^2 + \cdots,$$
$$q^4 = \tilde{q}^4 - 4\delta\tilde{q}^3 + \cdots.$$

Therefore the condition to eliminate the cubic term in U is $\mu_0 = 3\delta$.

(ii) The condition

$$\frac{\partial U}{\partial q} = 0$$

gives

$$q^3 + \mu_1 q + \mu_2 = 0. \qquad (3)$$

Equation (3) defines a surface in the (q, μ_1, μ_2) space. The solution of the cubic Eq. (3) is given by

$$q_1 = y_1 + y_2,$$

$$q_2 = -\frac{y_1 + y_2}{2} + \frac{y_1 - y_2}{2}i\sqrt{3},$$

$$q_3 = -\frac{y_1 + y_2}{2} - \frac{y_1 - y_2}{2}i\sqrt{3},$$

where

$$y_1 := \sqrt[3]{-\frac{\mu_2}{2} + \sqrt{D}}, \quad y_2 := \sqrt[3]{-\frac{\mu_2}{2} - \sqrt{D}}.$$

Here

$$D := \left(\frac{\mu_2}{2}\right)^2 + \left(\frac{\mu_1}{3}\right)^3$$

is the *discriminant*. Since μ_1 and μ_2 are real we find (a) one root is real and the two others are complex conjugate if $D > 0$, (b) all roots are real and at least two are equal if $D = 0$, (c) all roots are real and unequal if $D < 0$. We now have to investigate whether the real solutions lead to maxima, minima or points of inflexion for the potential (2). If $D > 0$, the potential (2) has one minimum at $q = 0$. There is no qualitative change of the potential $J(q) = q^4/4$. If $D < 0$, the potential (2) has two minima symmetric to $q = 0$. The point $q = 0$ is a local maximum of the potential U. In this case we have a qualitative change of the potential (2). Consequently,

$$D = 0$$

is a bifurcation line in the $\mu_1 - \mu_2$ plane.

Remark.   *This problem is the so-called* cusp catastrophe *in catastrophe theory.*

iii) The van der Waals equation can be written near the critical point as

$$(V - V_c)^3 = 0$$

or

$$V^3 - 3V_c V^2 + 3V_c^2 V - V_c^3 = 0 \tag{4}$$

where V_c is the critical pressure. This equation is compared with the van der Waals equation with $T = T_c$ and $P = P_c$, i.e.,

$$(V - b)\left(P_c + \frac{a}{V^2}\right) = RT_c. \tag{5}$$

From (5) it follows that

$$V^3 - \left(b + \frac{RT_c}{P_c}\right)V^2 + \frac{a}{P_c}V - \frac{ab}{P_c} = 0. \tag{6}$$

Comparing (4) and (6) we obtain

$$3V_c = b + \frac{RT_c}{P_c}, \quad 3V_c^2 = \frac{a}{P_c}, \quad V_c^3 = \frac{ab}{P_c}. \tag{7}$$

The solution of (7) is given by

$$RT_c = \frac{8a}{27b}, \quad P_c = \frac{a}{27b^2}, \quad V_c = 3b.$$

We now introduce the normalized quantities

$$\bar{P} := \frac{P}{P_c}, \quad \bar{T} := \frac{T}{T_c}, \quad \bar{V} := \frac{V}{V_c}.$$

We find

$$\left(\bar{P} + \frac{3}{\bar{V}^2}\right)\left(\bar{V} - \frac{1}{3}\right) = \frac{8}{3}\bar{T}. \tag{8}$$

Introducing the density

$$\bar{X} := \frac{1}{\bar{V}} \tag{9}$$

leads to

$$(\bar{P} + 3\bar{X}^2)\left(\frac{1}{\bar{X}} - \frac{1}{3}\right) = \frac{8}{3}\bar{T}. \tag{10}$$

With the transformation

$$p := \bar{P} - 1, \quad x := \bar{X} - 1, \quad t := \bar{T} - 1, \tag{11}$$

we obtain

$$x^3 + \frac{1}{3}(8t + p)x + \frac{1}{3}(8t - 2p) = 0. \tag{12}$$

Therefore

$$x^3 + \mu_1 x + \mu_2 = 0 \tag{13}$$

where

$$\mu_1 := \frac{1}{3}(8t + p), \quad \mu_2 := \frac{1}{3}(8t - 2p). \tag{14}$$

Problem 12.   Let $\mathbf{N}$ be the natural numbers. A set is called *denumerable* if it is equipotent to the set of natural numbers $\mathbf{N}$. Show that the set $\mathbf{N} \times \mathbf{N}$ is denumerable. This means find a $1 - 1$ map between $\mathbf{N}$ and $\mathbf{N} \times \mathbf{N}$.

Solution.   We write the elements of $\mathbf{N} \times \mathbf{N}$ in the form of an array as follows

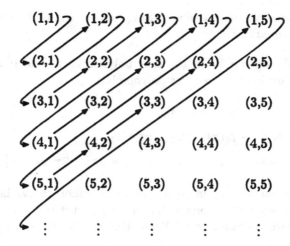

From the figure we see that we can arrange the elements of $\mathbf{N} \times \mathbf{N}$ into a (linear) sequence as indicated by the arrows, i.e.

$$(1,1), (2,1), (1,2), (3,1), (2,2), (1,3), (4,1), (3,2), \ldots$$

Thus we find a $1 - 1$ map between $\mathbf{N}$ and $\mathbf{N} \times \mathbf{N}$. If $(m,n) \in \mathbf{N} \times \mathbf{N}$ we have

$$f(m,n) = \frac{1}{2}(m + n - 1)(m + n - 2) + n \tag{1}$$

Equation (1) can be found as follows: The pair (m, n) lies in the $(m+n-1)$-th diagonal stripe in the above figure and is the n-th pair up from the left on that stripe. In the first stripe there is only one pair, in the second there are two pairs, and so on. Thus (m, n) occurs in position numbered

$$n + \sum_{k=1}^{m+n-2} k$$

in the counting procedure. Thus (1) follows.

How do you find m and n if $f(m, n)$ is given?

Problem 13. Consider the system of polynomial equations

$$f_1(\mathbf{x}) = x_1 - x_2 - x_3 = 0,$$
$$f_2(\mathbf{x}) = x_1 + x_2 - x_3^2 = 0,$$
$$f_3(\mathbf{x}) = x_1^2 + x_2^2 - 1 = 0.$$

Use a *Gaussian elimination-like method* to find the solutions of this system of polynomial equations.

Solution. We choose x_1 in the first polynomial as the first term for eliminating terms in the two other polynomials.

$$g_1(\mathbf{x}) = f_1(\mathbf{x}) = x_1 - x_2 - x_3 \,,$$

$$g_2(\mathbf{x}) = f_2(\mathbf{x}) - f_1(\mathbf{x}) = 2x_2 - x_3^2 + x_3 \,,$$

$$g_3(\mathbf{x}) = f_3(\mathbf{x}) - (x_1 + x_2 + x_3)f_1(\mathbf{x}) = 2x_2^2 + 2x_2 x_3 + x_3^2 - 1 \,.$$

We choose the variable x_2 in g_2 as the most important variable. Then we multiply g_2 by another polynomial and subtract it from $2g_2$ in order to eliminate the terms containing x_2. We do the same for the third polynomial

$$h_1(\mathbf{x}) = 2g_1(\mathbf{x}) + g_2(\mathbf{x}) = 2x_1 - x_3^2 - x_3 \,,$$

$$h_2(\mathbf{x}) = g_2(\mathbf{x}) = 2x_2 - x_3^2 + x_3 \,,$$

$$h_3(\mathbf{x}) = 2g_3(\mathbf{x}) - (2x_2 + x_3^2 + x_3)g_2(\mathbf{x}) = x_3^4 + x_3^2 - 2 \,.$$

The new set of equations is in upper triangular-form. The last polynomial is only in x_3, the second one is only in x_2 and x_3, and the first one is a polynomial in x_1 and x_3.

Problem 14. (i) Find the maximum area of all triangles that can be inscribed in an ellipse

$$\frac{x^2}{a^2} + \frac{y^2}{b^2} = 1$$

with semiaxes a and b.

(ii) Describe the triangles that have maximum area.

Solution (i) We represent the ellipse using the parametric representation

$$x(t) = a\cos(t)\,, \quad y(t) = b\sin(t)$$

where $t \in [0, 2\pi)$. A triple of points on the ellipse is given by

$$(a\cos t_j, b\sin t_j), j = 1, 2, 3 \,.$$

Thus the area A of an inscribed triangle is given by

$$A = \frac{1}{2} \begin{vmatrix} 1 & a\cos t_1 & b\sin t_1 \\ 1 & a\cos t_2 & b\sin t_2 \\ 1 & a\cos t_3 & b\sin t_3 \end{vmatrix} = \frac{ab}{2} \begin{vmatrix} 1 & \cos t_1 & \sin t_1 \\ 1 & \cos t_2 & \sin t_2 \\ 1 & \cos t_3 & \sin t_3 \end{vmatrix} .$$

This is ab times the area of a triangle inscribed in the unit circle. Hence, the area is maximal if

$$t_2 = t_1 + \frac{2\pi}{3} \quad \text{and} \quad t_3 = t_2 + \frac{2\pi}{3} .$$

The area A is given by

$$A = \frac{3\sqrt{3}}{4} ab$$

where we used the fact that $\sin(2\pi/3) = \sqrt{3}/2$ and $\cos(2\pi/3) = -1/2$.

(ii) Thus the area is maximal when the corresponding triangle inscribed in the unit circle is regular.

Chapter 9

L'Hospital's Rule

Problem 1. (i) Let

$$f(x) = x \ln(x) \quad \text{for } x > 0.$$

Determine $f(0)$ using *L'Hospital's rule*.

(ii) Let

$$f(\theta) = \frac{\cos(\frac{\pi}{2} \cos \theta)}{\sin \theta} \quad \text{for } \theta \neq n\pi$$

where $n \in \mathbf{Z}$. Determine $f(n\pi)$ using L'Hospital's rule.

Solution. (i) Since

$$\lim_{x \to +0} x \ln x \equiv \lim_{x \to +0} \frac{\ln x}{1/x},$$

we have $g(x) = \ln x$ and $h(x) = 1/x$. It follows that

$$\frac{dg}{dx} = \frac{1}{x}, \quad \frac{dh}{dx} = -\frac{1}{x^2}.$$

Therefore

$$\lim_{x \to +0} x \ln x = \lim_{x \to +0} \frac{\frac{1}{x}}{\frac{-1}{x^2}} = \lim_{x \to +0} \frac{-x^2}{x} = 0.$$

(ii) Let

$$g(\theta) = \cos\left(\frac{\pi}{2} \cos \theta\right), \quad h(\theta) = \sin \theta.$$

Then

$$\frac{dg}{d\theta} = \sin\left(\frac{\pi}{2}\cos\theta\right)\frac{\pi}{2}\sin\theta$$

and

$$\frac{dh}{d\theta} = \cos\theta.$$

Since

$$\left.\frac{dg}{d\theta}\right|_{\theta=n\pi} = 0$$

and

$$\left.\frac{dh}{d\theta}\right|_{\theta=n\pi} \neq 0,$$

we obtain

$$\lim_{\theta\to n\pi} f(\theta) = 0.$$

Problem 2. (i) Calculate

$$\lim_{\beta\to\infty} \tanh(\beta x)$$

where $x \in \mathbf{R}$.

(ii) Calculate

$$\lim_{\beta\to\infty} \frac{1}{\beta}\ln(4\cosh(\beta x)\cosh(\beta y))$$

where $x, y \in \mathbf{R}$.

(iii) Find

$$\lim_{\beta\to\infty} \frac{\sinh(\beta x)}{\cosh(\beta x) + \cosh(\beta y)}$$

where $\beta > 0$.

Solution.   (i) Let $x = 0$, then

$$\tanh(0) = 0.$$

If $x > 0$, then

$$\lim_{\beta\to\infty} \tanh(\beta x) = 1.$$

If $x < 0$, then

$$\lim_{\beta \to \infty} \tanh(\beta x) = -1.$$

Thus

$$\lim_{\beta \to \infty} \tanh(\beta x) = \begin{cases} 1 & x > 0, \\ 0 & x = 0, \\ -1 & x < 0. \end{cases}$$

One writes

$$\lim_{\beta \to \infty} \tanh(\beta x) = \operatorname{sgn}(x). \tag{1}$$

(ii) Applying L'Hospital's rule we have

$$\lim_{\beta \to \infty} \frac{1}{\beta} \ln[4 \cosh(\beta x) \cosh(\beta y)]$$

$$= \lim_{\beta \to \infty} \frac{x \sinh(\beta x) \cosh(\beta y) + y \cosh(\beta x) \sinh(\beta y)}{\cosh(\beta x) \cosh(\beta y)}$$

$$= \lim_{\beta \to \infty} [x \tanh(\beta x) + y \tanh(\beta y)] = x \operatorname{sgn}(x) + y \operatorname{sgn}(y) = |x| + |y|.$$

(iii) Applying the addition theorem and (1) we find

$$\lim_{\beta \to \infty} \frac{\sinh(\beta x)}{\cosh(\beta x) + \cosh(\beta y)}$$

$$= \lim_{\beta \to \infty} \frac{\sinh[\beta((x/2 + y/2) + (x/2 - y/2))]}{\cosh(\beta x) + \cosh(\beta y)}$$

$$= \frac{1}{2} \lim_{\beta \to \infty} \tanh\left(\beta \frac{x + y}{2}\right) + \frac{1}{2} \lim_{\beta \to \infty} \tanh\left(\beta \frac{x - y}{2}\right)$$

$$= \frac{1}{2} \operatorname{sgn}(x + y) + \frac{1}{2} \operatorname{sgn}(x - y)$$

where we have used the identity

$$\frac{\sinh[\beta((x + y/2) + (x - y/2))]}{\cosh(\beta x) + \cosh(\beta y)}$$

$$\equiv \frac{\sinh(\beta x + y/2) \cosh(\beta x - y/2) + \cosh(\beta x + y/2) \sinh(\beta x - y/2)}{2 \cosh(\beta x + y/2) \cosh(\beta x - y/2)}.$$

Chapter 10

Lagrange Multiplier Method

Problem 1. Let M be a manifold and f be a real valued function of class $C^{(1)}$ on some open set containing M. We consider the problem of finding the extrema of the function $f|M$. This is called a problem of *constrained extrema*.

The *Lagrange multiplier rule* is as follows. Assume that f has a constrained relative extremum at $\mathbf{x}^* = (x_1^*, x_2^*, \ldots, x_n^*)$. Then there exist real numbers $\lambda_1, \lambda_2, \ldots, \lambda_m$ such that $\mathbf{x}^*$ is a critical point of the function

$$F(\mathbf{x}) := f(\mathbf{x}) + \lambda_1 g_1(\mathbf{x}) + \cdots + \lambda_m g_m(\mathbf{x}).$$

The numbers $\lambda_1, \lambda_2, \ldots, \lambda_m$ are called *Lagrange multipliers*. The critical points are given by the solution of the system

$$\frac{\partial F}{\partial x_k} = 0, \quad k = 1, 2, \ldots, n.$$

Let

$$f(\mathbf{x}) = x_1 - x_2 + 2x_3.$$

Find the maximum and minimum values of f on the ellipsoid

$$M := \{(x_1, x_2, x_3) : x_1^2 + x_2^2 + 2x_3^2 = 2\}.$$

Solution. Let

$$g(\mathbf{x}) = 2 - (x_1^2 + x_2^2 + 2x_3^2)$$

and

$$F(\mathbf{x}) = f(\mathbf{x}) + \lambda g(\mathbf{x}).$$

The Lagrange multiplier λ is yet to be determined. From the multiplier rule we obtain three equations

$$\left. \frac{\partial F}{\partial x_1} \right|_{\mathbf{x} = \mathbf{x}^*} = 1 - 2\lambda x_1^* = 0,$$

$$\left. \frac{\partial F}{\partial x_2} \right|_{\mathbf{x} = \mathbf{x}^*} = -1 - 2\lambda x_2^* = 0,$$

$$\left. \frac{\partial F}{\partial x_3} \right|_{\mathbf{x} = \mathbf{x}^*} = 2 - 4\lambda x_3^* = 0.$$

From these and the fourth equation (constraint) $g(\mathbf{x}) = 0$, we obtain

$$x_1^* = \frac{1}{2\lambda}, \quad x_2^* = -\frac{1}{2\lambda}, \quad x_3^* = \frac{1}{2\lambda}, \quad \frac{1}{\lambda} = \pm\sqrt{2}.$$

Therefore

$$\mathbf{x}^* = \pm(\sqrt{2}/2)(\mathbf{e}_1 - \mathbf{e}_2 + \mathbf{e}_3)$$

depending on which of the two possible values for λ is used. Here $\mathbf{e}_1$, $\mathbf{e}_2$ and $\mathbf{e}_3$ denote the standard basis in $\mathbf{R}^3$. Since f is continuous and M is a compact set, f has a maximum and a minimum value on M. One of the two critical points obtained by the multiplier rule must give the maximum and the other the minimum. Since

$$f[(\sqrt{2}/2)(\mathbf{e}_1 - \mathbf{e}_2 + \mathbf{e}_3)] = 2\sqrt{2}$$

and

$$f[-(\sqrt{2}/2)(\mathbf{e}_1 - \mathbf{e}_2 + \mathbf{e}_3)] = -2\sqrt{2},$$

these numbers are the maximum and minimum values, respectively.

Problem 2.   Calculate the greatest volume of a rectangular box that can be securely tied up with a piece of string 360 cm in length. The string must pass twice around the width of the parcel and once around the length as shown in the figure. Disregard the amount used up by the knot.

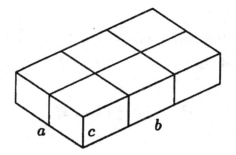

Solution. The problem can be solved using the Lagrange multiplier method. Let a, b, c be the width, length and height respectively of the box. The length of the string is

$$S = 2b + 4a + 6c = 360 \text{ cm}. \tag{1}$$

This is the constraint. The volume of the box is $V = abc$. Thus we have

$$F(a, b, c) = abc + \lambda(2b + 4a + 6c - 360)$$

where λ is the Lagrange multiplier. Therefore

$$\frac{\partial F}{\partial a} = bc + 4\lambda = 0, \tag{2a}$$

$$\frac{\partial F}{\partial b} = ac + 2\lambda = 0, \tag{2b}$$

$$\frac{\partial F}{\partial c} = ab + 6\lambda = 0. \tag{2c}$$

From (2a) through (2c) we obtain

$$2a^* = b^*, \quad b^* = 3c^*, \quad 2a^* = 3c^*. \tag{3}$$

Inserting (3) into the constraint (1) yields

$$a^* = 30 \text{ cm}, \quad b^* = 60 \text{ cm}, \quad c^* = 20 \text{ cm}.$$

This is obviously a global maximum.

Problem 3. Let

$$f(\mathbf{x}) = x_1 x_2 \cdots x_n$$

and

$$M := \{(x_1, x_2, \ldots, x_n) : x_1^2 + x_2^2 + \cdots + x_n^2 = 1\}.$$

Find the extrema of $f|_M$.

Solution.   Let

$$r^2 := x_1^2 + x_2^2 + \cdots + x_n^2.$$

From the system

$$\frac{\partial}{\partial x_k}(f(\mathbf{x}) + \lambda(1 - r^2)) = 0, \quad k = 1, 2, \ldots, n$$

we find

$$2\lambda x_k^{*2} = x_1^* x_2^* \cdots x_n^*, \quad k = 1, 2, \ldots, n \tag{1}$$

and

$$2\lambda = n x_1^* x_2^* \cdots x_n^* \tag{2}$$

where we have taken into account the constraint $r^2 = 1$. From (1) and (2) we obtain

$$x_k^{*2} = \frac{1}{n}, \quad k = 1, 2, \ldots, n.$$

Thus at the extrema we have

$$f(\mathbf{x}^*) = \pm\sqrt{\frac{1}{n^n}}.$$

Chapter 11

Linear Difference Equations

Problem 1. Let $n, m \in \mathbf{N}$ with $n > m$. Show that

$$\sum_{k=m}^{n} u_k(v_{k+1} - v_k) \equiv u_k v_k \big|_m^{n+1} - \sum_{k=m}^{n} v_{k+1}(u_{k+1} - u_k) \qquad (1)$$

where

$$u_k v_k \big|_m^{n+1} = u_{n+1} v_{n+1} - u_m v_m .$$

Remark. *Equation (1) is called the formula for summation by parts.*

Solution. We start from the identity

$$u_{k+1} v_{k+1} - u_k v_k \equiv u_k(v_{k+1} - v_k) + v_{k+1}(u_{k+1} - u_k) .$$

Therefore

$$\sum_{k=m}^{n} (u_{k+1} v_{k+1} - u_k v_k) \equiv \sum_{k=m}^{n} u_k(v_{k+1} - v_k) + \sum_{k=m}^{n} v_{k+1}(u_{k+1} - u_k) .$$

Obviously, the left-hand side is given by

$$\sum_{k=m}^{n} (u_{k+1} v_{k+1} - u_k v_k) \equiv u_{n+1} v_{n+1} - u_m v_m .$$

Problem 2. (i) Give the solution of the linear difference equation

$$x_{t+1} = 2x_t \qquad (1)$$

where $t = 0, 1, 2, \ldots$ and $x_0 = 1$.

(ii) Prove that the set of n (different) elements has exactly 2^n (different) subsets.

Solution.   (i) Since (1) is a linear difference equation with constant coefficients we can solve the equation with the ansatz

$$x_t = ar^t$$

where a is a constant. Inserting this ansatz into (1) gives $r = 2$. Therefore the solution of (1) is given by

$$x_t = a2^t .$$

Inserting the initial condition $x_0 = 1$ leads to $a = 1$ and therefore

$$x_t = 2^t .$$

(ii) For each n, let X_n denote the number of (different) subsets of a set with n (different) elements. Let S be a set with $n+1$ elements, and designate one of its elements by s. There is a one-to-one correspondence between those subsets of S which do not contain s and those subsets that do contain s (namely, a subset T of the former type corresponds to $T \cup \{s\}$). The former types are all subsets of $S \setminus \{s\}$, a set with n elements, and therefore, it must be the case that

$$X_{n+1} = 2X_n$$

where $X_0 = 1$. Using the result from (i) we obtain

$$X_n = 2^n .$$

Problem 3.   (i) Find the solution of the linear difference equation with constant coefficients

$$x_{t+2} - x_{t+1} - x_t = 0 \tag{1}$$

where $t = 0, 1, 2, \ldots$ and $x_0 = 0$, $x_1 = 1$.

Remark.   *The numbers x_t are called* Fibonacci numbers.

(ii) *Find*

$$g = \lim_{t \to \infty} \frac{x_{t+1}}{x_t} . \tag{2}$$

Solution. (i) Since (1) is a linear difference equation with constant coefficients we can solve (1) with the ansatz

$$x_t = ar^t \tag{3}$$

where a is a constant. Inserting ansatz (3) into (1) yields

$$r^2 - r - 1 = 0. \tag{4}$$

The solution to (4) is given by

$$r_{1,2} = \frac{1}{2} \pm \sqrt{\frac{5}{4}} = \frac{1}{2}(1 \pm \sqrt{5}).$$

Consequently, the solution to (1) is given by

$$x_t = a_1 r_1^t + a_2 r_2^t$$

where a_1 and a_2 are the two constants of "integration". Imposing the initial condition $x_0 = 0$, $x_1 = 1$ yields

$$x_t = \frac{1}{\sqrt{5}} \left[\left(\frac{1}{2}(1 + \sqrt{5}) \right)^t - \left(\frac{1}{2}(1 - \sqrt{5}) \right)^t \right].$$

ii) From (2) we find

$$g = \lim_{t \to \infty} \frac{x_{t+1}}{x_t} = \lim_{t \to \infty} \frac{x_{t+2}}{x_{t+1}} = \lim_{t \to \infty} \frac{x_{t+1} + x_t}{x_{t+1}} = 1 + \lim_{t \to \infty} \frac{x_t}{x_{t+1}}$$

$$= 1 + \lim_{t \to \infty} \frac{1}{\frac{x_{t+1}}{x_t}}.$$

Thus g is the solution of the equation $g = 1 + 1/g$ with $g > 1$.

Problem 4. A coin is tossed n times. What is the probability that two heads will turn up in succession somewhere in the sequence of throws?

Solution. Let P_n denote the probability that two consecutive heads do not appear in n throws. Obviously

$$P_1 = 1, \quad P_2 = \frac{3}{4}.$$

If $n > 2$, there are two cases. If the first throw is tails, then two consecutive heads will not appear in the remaining $n - 1$ tosses with probability P_{n-1} (by our choice of notation). If the first throw is heads, the second toss must be tails to avoid two consecutive heads, and then two consecutive heads will not appear in the remaining $n-2$ throws with probability P_{n-2}. Thus, we obtain the linear difference equation

$$P_n = \frac{1}{2}P_{n-1} + \frac{1}{4}P_{n-2}, \quad n > 2.$$

This difference equation can be transformed to a more familiar form by multiplying each side by 2^n

$$2^n P_n = 2^{n-1}P_{n-1} + 2^{n-2}P_{n-2}.$$

Defining

$$S_n := 2^n P_n$$

we obtain

$$S_n = S_{n-1} + S_{n-2}.$$

This is the linear difference equation for the Fibonacci sequence. Note that

$$S_n = F_{n+2}.$$

Thus, the probability is given by

$$Q_n = 1 - P_n = 1 - \frac{F_{n+2}}{2^n}.$$

Problem 5.   (i) Let A be an $n \times n$ matrix with constant entries. Then

$$\mathbf{x}_{t+1} = A\mathbf{x}_t \tag{1}$$

is a system of n linear difference equations with constant coefficients, where $t = 0, 1, 2, \ldots$. Show that the solution of the initial-value problem is given by

$$\mathbf{x}_t = A^t\mathbf{x}_0 \tag{2}$$

where $\mathbf{x}_0$ is the initial vector.

(ii) Let

$$A = \begin{pmatrix} 1 & 0 & 0 & 0 \\ 1 & 1 & 0 & 0 \\ 0 & 1 & 1 & 0 \\ -1 & -1 & 0 & 1 \end{pmatrix}.$$

Find the solution of the difference Eq. (1).

Solution.   (i) From (2) it follows that

$$x_{t+1} = A^{t+1}x_0 = A^t A x_0 = A A^t x_0 = A x_t .$$

(ii) We set

$$A = I + B$$

where I is the 4×4 unit matrix. Therefore

$$B = \begin{pmatrix} 0 & 0 & 0 & 0 \\ 1 & 0 & 0 & 0 \\ 0 & 1 & 0 & 0 \\ -1 & -1 & 0 & 0 \end{pmatrix} .$$

Now

$$A^t = (I + B)^t = I^t + t I^{t-1} B + \frac{t(t-1)}{2} I^{t-2} B^2 + \cdots + B^t$$

since $[I, B] = 0$. Thus

$$A^t = I + tB + \frac{t(t-1)}{2} B^2 + \cdots + B^t .$$

Since

$$B^2 = BB = \begin{pmatrix} 0 & 0 & 0 & 0 \\ 0 & 0 & 0 & 0 \\ 1 & 0 & 0 & 0 \\ -1 & 0 & 0 & 0 \end{pmatrix}$$

and

$$B^3 = B^2 B = \begin{pmatrix} 0 & 0 & 0 & 0 \\ 0 & 0 & 0 & 0 \\ 0 & 0 & 0 & 0 \\ 0 & 0 & 0 & 0 \end{pmatrix} ,$$

we obtain

$$A^t = I + tB + \frac{t(t-1)}{2} B^2 .$$

Therefore the solution is given by

$$\mathbf{x}_t = \left(I + tB + \frac{t(t-1)}{2}B^2\right)\mathbf{x}_0.$$

Problem 6.   Sum the finite series

$$S := a_0 + a_1 + \cdots + a_T$$

where $a_0 = 2$ and $a_1 = 5$ for

$$a_{t+2} = 5a_{t+1} - 6a_t \tag{1}$$

with $t = 0, 1, 2, \ldots$.

Solution.   The first few terms of the a_t-sequence are

$$2, 5, 13, 35, 97, 275, 793, \ldots.$$

A general formula for the nth term is not apparent. We apply the technique of *generating functions*. Let

$$F(x) := a_0 + a_1 x + a_2 x^2 + \cdots + a_t x^t + \cdots \tag{2}$$

be the ansatz for the generating function. We obtain from (2)

$$-5xF(x) = -5a_0 x - 5a_1 x^2 - \cdots - 5a_{t+1} x^{t+2} - \cdots, \tag{3a}$$

$$6x^2 F(x) = 6a_0 x^2 + \cdots + 6a_t x^{t+2} + \cdots. \tag{3b}$$

Adding (2), (3a) and (3b) and using (1), we obtain

$$(1 - 5x + 6x^2)F(x) = a_0 + (a_1 - 5a_0)x,$$

so that

$$F(x) = \frac{2 - 5x}{(1 - 2x)(1 - 3x)}.$$

We write this as a sum of partial fractions

$$F(x) = \frac{1}{1 - 2x} + \frac{1}{1 - 3x}.$$

Making use of the geometric series we have

$$F(x) = \sum_{t=0}^{\infty}(2x)^t + \sum_{t=0}^{\infty}(3x)^t \equiv \sum_{t=0}^{\infty}(2^t + 3^t)x^t.$$

Thus

$$a_t = 2^t + 3^t$$

for $t = 0, 1, 2, \ldots$. Consequently, the sum is given by

$$S = a_0 + a_1 + \cdots + a_T = \sum_{t=0}^{T}(2^t + 3^t) = \sum_{t=0}^{T} 2^t + \sum_{t=0}^{T} 3^t \,.$$

Thus

$$S = \frac{2^{T+1} - 1}{2 - 1} + \frac{3^{T+1} - 1}{3 - 1} = 2^{T+1} - 1 + \frac{3^{T+1} - 1}{2} = \frac{2^{T+2} + 3^{T+1} - 3}{2} \,.$$

Remark. *The problem can also be solved by solving the linear difference equation with constant coefficients* (1) *using the ansatz*

$$a_t \propto r^t$$

and the initial conditions $a_0 = 2$ and $a_1 = 5$.

Problem 7. An electrical network has the form shown in the following figure, where R_1 and R_2 are given constant resistances. V is the voltage applied by a source.

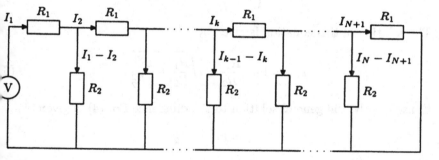

i Write down *Kirchhoff's law* for the first loop, the kth loop and the $(N + 1)$th loop.

ii) Give the general solution for the difference equation of the kth loop.

iii) Determine the constants in the general solution of the difference equation with the boundary conditions (first and $(N + 1)$th loop).

Solution. (i) For the first loop which includes the applied voltage Kirchhoff's law provides us with

$$I_1 R_1 + (I_1 - I_2)R_2 - V = 0 \,. \tag{1}$$

For the kth loop we find

$$I_k R_1 + (I_k - I_{k+1})R_2 - (I_{k-1} - I_k)R_2 = 0 . \qquad (2)$$

Kirchhoff's law for the last loop $((N+1)$th loop) gives

$$I_{N+1}R_1 - (I_N - I_{N+1})R_2 = 0 . \qquad (3)$$

Equation (2) can be written as the difference equation

$$I_{k+1} - \left(2 + \frac{R_1}{R_2}\right) I_k + I_{k-1} = 0 \qquad (4)$$

where $k = 1, 2, \ldots, N$.

(ii) Equation (4) is a linear difference equation with constant coefficients. It can therefore be solved with the ansatz

$$I_k = ar^k . \qquad (5)$$

The first and last loop provide the boundary conditions. Inserting the ansatz (5) into (4) leads to the algebraic equation

$$r^2 - \left(2 + \frac{R_1}{R_2}\right) r + 1 = 0 .$$

The solution of this equation is given by

$$r_{1,2} = 1 + \frac{R_1}{2R_2} \pm \sqrt{\frac{R_1}{R_2} + \frac{R_1^2}{4R_2^2}} . \qquad (6)$$

Consequently, the general solution to the difference Eq. (4) is given by

$$I_k = ar_1^k + br_2^k$$

where a and b are two constants.

(iii) Now we have to impose the boundary conditions (1) and (3). From (1) we find

$$I_2 = \frac{I_1(R_1 + R_2) - V}{R_2} \qquad (7)$$

and from (3) we obtain

$$I_N = \frac{(R_1 + R_2)I_{N+1}}{R_2} . \qquad (8)$$

If we set $k = 2$ and $k = N$ in (8) we arrive at

$$I_2 = ar_1^2 + br_2^2, \quad I_N = ar_1^N + br_2^N.$$

It follows that

$$a = \frac{I_2 r_2^N - I_N r_2^2}{r_1^2 r_2^N - r_2^2 r_1^N}, \quad b = \frac{I_N r_1^2 - I_2 r_1^N}{r_1^2 r_2^N - r_2^2 r_1^N}.$$

Therefore

$$I_k = \left(\frac{I_2 r_2^N - I_N r_2^2}{r_1^2 r_2^N - r_2^2 r_1^N}\right) r_1^k + \left(\frac{I_N r_1^2 - I_2 r_1^N}{r_1^2 r_2^N - r_2^2 r_1^N}\right) r_2^k$$

where r_1, r_2 are given in (6) and the currents I_2, I_N are given by (7) and (8).

Problem 8. Consider the coupled system of linear difference equations

$$W_{r+1}(k, l, 1) = W_r(k-1, l, 1) + e^{-i\pi/4} W_r(k, l-1, 2) + e^{i\pi/4} W_r(k, l+1, 4),$$
(1a)

$$W_{r+1}(k, l, 2) = e^{i\pi/4} W_r(k-1, l, 1) + W_r(k, l-1, 2) + e^{-i\pi/4} W_r(k+1, l, 3),$$
(1b)

$$W_{r+1}(k, l, 3) = e^{i\pi/4} W_r(k, l-1, 2) + W_r(k+1, l, 3) + e^{-i\pi/4} W_r(k, l+1, 4),$$
(1c)

$$W_{r+1}(k, l, 4) = e^{-i\pi/4} W_r(k-1, l, 1) + e^{i\pi/4} W_r(k+1, l, 3) + W_r(k, l+1, 4),$$
(1d)

where $k, l \in \{0, 1, 2, \ldots, L-1\}$ and $L \equiv 0$ (periodic boundary conditions). Equations (1a) through (1d) can be written in the matrix form

$$W_{r+1}(k, l, \nu) = \sum_{k', l', \nu'} M(kl\nu, k'l'\nu') W_r(k', l', \nu')$$

where $\nu = 1, 2, 3, 4$. Diagonalize M with respect to k and l by applying the discrete Fourier transform.

Solution. The *discrete Fourier transform* in two dimensions is given by

$$\hat{W}_r(p, q, \nu) = \sum_{k=0}^{L-1} \sum_{l=0}^{L-1} e^{-2\pi i(pk+ql)/L} W_r(k, l, \nu).$$

The inverse Fourier transformation follows as

$$W_r(k, l, \nu) = \frac{1}{L^2} \sum_{p=0}^{L-1} \sum_{q=0}^{L-1} e^{2\pi i(pk+ql)/L} \hat{W}_r(p, q, \nu) \,.$$

Taking Fourier components on both sides of (1) we find that each equation contains only $\hat{W}_r(p, q, \nu)$ with the same p, q. Consequently, the matrix M is diagonal with respect to p and q. For given p, q we find the 4×4 matrix

$$M(pq\nu|pq\nu') = \begin{pmatrix} \beta^{-p} & \alpha^{-1}\beta^{-q} & 0 & \alpha\beta^q \\ \alpha\beta^{-p} & \beta^{-q} & \alpha^{-1}\beta^p & 0 \\ 0 & \alpha\beta^{-q} & \beta^p & \alpha^{-1}\beta^q \\ \alpha^{-1}\beta^{-p} & 0 & \alpha\beta^p & \beta^q \end{pmatrix}$$

where $\alpha := e^{i\pi/4}$ and $\beta := e^{2\pi i/L}$.

Problem 9.   The *Stirling number* of the second kind $S(n, m)$ is the number of partitions of a set with n elements into m classes. For example, let

$$X := \{a, b, c, d\} \,. \tag{1}$$

Thus $n = 4$. Let $m = 2$. Then the partitions are

$$\{\{a\}, \{b, c, d\}\}; \{\{b\}, \{a, c, d\}\}; \{\{c\}, \{a, b, d\}\}; \{\{d\}, \{a, b, c\}\};$$

$$\{\{a, b\}, \{c, d\}\}; \{\{a, c\}, \{b, d\}\}; \{\{a, d\}, \{b, c\}\}\} \,.$$

Thus we have seven partitions.
Show that $S(n, m)$ satisfies the linear difference equation

$$S(n + 1, m) = S(n, m - 1) + mS(n, m) \tag{2a}$$

with the initial condition

$$S(n, 1) = S(n, n) = 1 \,. \tag{2b}$$

Solution.   Consider a set with n elements, i.e.

$$X := \{x_1, x_2, \ldots, x_n\} \,.$$

Let $S(n, m - 1)$ be the number of partitions into $m - 1$ classes. One can obtain $S(n, m - 1)$ partitions into m classes of a set with $n + 1$ elements $x_1, \ldots, x_{n+1}$ by adding to each partition a new class consisting of only the element x_{n+1}. The element x_{n+1} can be added to each of the already existing m classes of a partition of X in m distinct ways. These two procedures yield, without repetitions, all the partitions of the set X into m classes. Equation (2) follows.

Remark. *The solution to (1) is given by*

$$S(n,m) = \frac{1}{m!} \sum_{k=0}^{m-1} (-1)^k \binom{m}{k} (m-k)^n.$$

Chapter 12

Linear Differential Equations

Problem 1. Find the general solution for the *driven damped harmonic oscillator*

$$\frac{d^2u}{dt^2} + \alpha\frac{du}{dt} + \omega^2 u = k_1\cos(\Omega_1 t) + k_2\cos(\Omega_2 t) \tag{1}$$

where $\alpha > 0$ and $\Omega_1 \neq \Omega_2$. Discuss the dependence on Ω_1 and Ω_2.

Solution. The general solution of Eq. (1) is given by

$$u = u_h + u_p$$

where u_h is the general solution to the homogeneous equation and u_p is a particular solution to the inhomogeneous equation. We first determine the general solution of the homogeneous differential equation

$$\frac{d^2u}{dt^2} + \alpha\frac{du}{dt} + \omega^2 u = 0. \tag{2}$$

Since (2) is a linear differential equation with constant coefficients we can solve the equation with the exponential ansatz

$$u(t) \propto e^{\lambda t}.$$

Inserting this ansatz into (2) yields

$$\lambda^2 + \alpha\lambda + \omega^2 = 0.$$

The roots are

$$\lambda_\pm = -\frac{\alpha}{2} \pm \sqrt{\left(\frac{\alpha}{2}\right)^2 - \omega^2}.$$

Consequently, the general solution of the homogeneous part of the differential equation is given by one of the following cases:

$$\omega^2 = \left(\frac{\alpha}{2}\right)^2 : u_h(t) = (C + Dt)e^{-\alpha t/2}, \tag{3a}$$

$$\omega^2 < \left(\frac{\alpha}{2}\right)^2 : u_h(t) = Ce^{(-\alpha/2+\sqrt{(\alpha/2)^2-\omega^2})t} + De^{(-\alpha/2-\sqrt{(\alpha/2)^2-\omega^2})t}, \tag{3b}$$

$$\omega^2 > \left(\frac{\alpha}{2}\right)^2 : u_h(t) = (C\cos\sqrt{\omega^2 - (\alpha/2)^2}t + D\sin\sqrt{\omega^2 - (\alpha/2)^2}t)e^{-\alpha t/2}, \tag{3c}$$

where C and D are the constants of integration. To find the particular solution to (1) we consider first the equation

$$\frac{d^2u}{dt^2} + \alpha\frac{du}{dt} + \omega^2 u = k\cos(\Omega t). \tag{4}$$

For this equation we make the ansatz

$$u_p(t) = A\cos(\Omega t) + B\sin(\Omega t)$$

where A and B are two constants to be determined. Inserting this ansatz into the differential Eq. (4) gives

$$(\omega^2 - \Omega^2)A + \alpha\Omega B = k, \quad -\alpha\Omega A + (\omega^2 - \Omega^2)B = 0. \tag{5}$$

Solving (5) for A and B we find

$$A = \frac{(\omega^2 - \Omega^2)k}{(\omega^2 - \Omega^2)^2 + \alpha^2\Omega^2}, \quad B = \frac{\alpha\Omega k}{(\omega^2 - \Omega^2)^2 + \alpha^2\Omega^2}.$$

Generalizing the right-hand side of (4) to two external forces, we make the ansatz

$$u_p(t) = A_1\cos(\Omega_1 t) + B_1\sin(\Omega_1 t) + A_2\cos(\Omega_2 t) + B_2\sin(\Omega_2 t).$$

We find the following linear equations for A_1, A_2, B_1 and B_2

$$k_1 = -A_1\Omega_1^2 + \alpha B_1\Omega_1 + \omega^2 A_1,$$
$$0 = -B_1\Omega_1^2 - \alpha A_1\Omega_1 + \omega^2 B_1,$$
$$k_2 = -A_2\Omega_2^2 + \alpha B_2\Omega_2 + \omega^2 A_2,$$
$$0 = -B_2\Omega_2^2 - \alpha A_2\Omega_2 + \omega^2 B_2.$$

Consequently, we find that the general solution to (1) is given by

$$u(t) = u_h(t) + \sum_{j=1}^{2} \frac{k_j}{(\omega^2 - \Omega_j^2)^2 + \alpha^2 \Omega_j^2}$$

$$\times [(\omega^2 - \Omega_j^2) \cos(\Omega_j t) + \alpha \Omega_j \sin(\Omega_j t)]$$

where u_h is given by (3).

Problem 2. The *motion of a charge q* in an electromagnetic field is given by

$$m \frac{d\mathbf{v}}{dt} = q(\mathbf{E} + \mathbf{v} \times \mathbf{B}) \tag{1}$$

where m denotes the mass and $\mathbf{v}$ the velocity. Assume that

$$\mathbf{E} = \begin{pmatrix} E_1 \\ E_2 \\ E_3 \end{pmatrix}, \quad \mathbf{B} = \begin{pmatrix} B_1 \\ B_2 \\ B_3 \end{pmatrix}$$

are constant fields. Find the solution of the initial value problem.

Solution. Equation (1) can be written in the form

$$\begin{pmatrix} dv_1/dt \\ dv_2/dt \\ dv_3/dt \end{pmatrix} = \frac{q}{m} \begin{pmatrix} 0 & B_3 & -B_2 \\ -B_3 & 0 & B_1 \\ B_2 & -B_1 & 0 \end{pmatrix} \begin{pmatrix} v_1 \\ v_2 \\ v_3 \end{pmatrix} + \frac{q}{m} \begin{pmatrix} E_1 \\ E_2 \\ E_3 \end{pmatrix}.$$

We set

$$B_j \to \frac{q}{m} B_j, \quad E_j \to \frac{q}{m} E_j.$$

Thus

$$\begin{pmatrix} dv_1/dt \\ dv_2/dt \\ dv_3/dt \end{pmatrix} = \begin{pmatrix} 0 & B_3 & -B_2 \\ -B_3 & 0 & B_1 \\ B_2 & -B_1 & 0 \end{pmatrix} \begin{pmatrix} v_1 \\ v_2 \\ v_3 \end{pmatrix} + \begin{pmatrix} E_1 \\ E_2 \\ E_3 \end{pmatrix}. \tag{2}$$

Equation (2) is a system of nonhomogeneous linear differential equations with constant coefficients. The solution to the homogeneous equation

$$\begin{pmatrix} dv_1/dt \\ dv_2/dt \\ dv_3/dt \end{pmatrix} = \begin{pmatrix} 0 & B_3 & -B_2 \\ -B_3 & 0 & B_1 \\ B_2 & -B_1 & 0 \end{pmatrix} \begin{pmatrix} v_1 \\ v_2 \\ v_3 \end{pmatrix} \tag{3}$$

s given by

$$\begin{pmatrix} v_1(t) \\ v_2(t) \\ v_3(t) \end{pmatrix} = e^{tM} \begin{pmatrix} v_1(0) \\ v_2(0) \\ v_3(0) \end{pmatrix}$$

where M is the matrix of the right hand side of (6) and $v_j(0) = v_j(t = 0)$. The solution of the system of nonhomogeneous linear differential Eq. (5) an be found with the help of the method called *variation of constants*. One sets

$$\mathbf{v}(t) = e^{tM} \mathbf{f}(t)$$

where $\mathbf{f} : \mathbf{R} \to \mathbf{R}^3$ is some differentiable curve. Then

$$\frac{d\mathbf{v}}{dt} = M e^{tM} \mathbf{f}(t) + e^{tM} \frac{d\mathbf{f}}{dt}. \tag{4}$$

inserting (4) into (2) yields

$$M\mathbf{v}(t) + \mathbf{E} = M e^{tM} \mathbf{f}(t) + e^{tM} \frac{d\mathbf{f}}{dt} = M\mathbf{v}(t) + e^{tM} \frac{d\mathbf{f}}{dt}.$$

onsequently

$$\frac{d\mathbf{f}}{dt} = e^{-tM} \mathbf{E}.$$

y integration we obtain

$$\mathbf{f}(t) = \int_0^t e^{-sM} \mathbf{E} ds + \mathbf{K}$$

here

$$\mathbf{K} = \begin{pmatrix} K_1 \\ K_2 \\ K_3 \end{pmatrix}.$$

herefore we obtain the general solution of the initial value problem of the nhomogeneous system (2), namely

$$\mathbf{v}(t) = e^{tM} \left(\int_0^t e^{-sM} \mathbf{E} ds + \mathbf{K} \right)$$

where

$$\mathbf{K} = \begin{pmatrix} v_1(0) \\ v_2(0) \\ v_3(0) \end{pmatrix}.$$

We now have to calculate e^{tM} and e^{-sM}. We find that

$$M^2 = MM = \begin{pmatrix} -B_2^2 - B_3^2 & B_1 B_2 & B_1 B_3 \\ B_1 B_2 & -B_1^2 - B_3^2 & B_2 B_3 \\ B_1 B_3 & B_2 B_3 & -B_1^2 - B_2^2 \end{pmatrix}$$

and

$$M^3 = M^2 M = -B^2 M \tag{5}$$

where

$$B^2 = B_1^2 + B_2^2 + B_3^2$$

and

$$B := \sqrt{B_1^2 + B_2^2 + B_3^2}.$$

Therefore

$$M^4 = -B^2 M^2.$$

Since

$$e^{tM} := \sum_{k=0}^{\infty} \frac{(tM)^k}{k!} = I + \frac{tM}{1!} + \frac{t^2 M^2}{2!} + \frac{t^3 M^3}{3!} + \frac{t^4 M^4}{4!} + \frac{t^5 M^5}{5!} + \cdots$$

where I denotes the 3×3 unit matrix, we obtain, taking (5) into account

$$e^{tM} = I + M \left(t - \frac{t^3}{3!} B^2 + \frac{t^5}{5!} B^4 - \cdots \right) + M^2 \left(\frac{t^2}{2!} - \frac{t^4}{4!} B^2 + \frac{t^6}{6!} B^4 - \cdots \right)$$

Thus

$$e^{tM} = I + \frac{M}{B} \left(tB - \frac{t^3}{3!} BB^2 + \cdots \right) - \frac{M^2}{B^2} \left(1 - 1 - \frac{t^2 B^2}{2!} + \frac{t^4 B^4}{4!} - \cdots \right)$$

$$= I + \frac{M}{B} \sin(Bt) + \frac{M^2}{B^2} (1 - \cos(Bt)).$$

Therefore

$$e^{tM} = I + \frac{M}{B}\sin(Bt) + \frac{M^2}{B^2}(1 - \cos(Bt)) \tag{6}$$

and

$$e^{-sM} = I - \frac{M}{B}\sin(Bs) + \frac{M^2}{B^2}(1 - \cos(Bs)). \tag{7}$$

Equation (7) follows from (6) by replacing $t \to -s$. Since

$$\int_0^t e^{-sM}\mathbf{E}\,ds = \int_0^t \left(I - \frac{M}{B}\sin(Bs) + \frac{M^2}{B^2}(1 - \cos(Bs))\right)\mathbf{E}\,ds$$

$$= \mathbf{E}t + \frac{M\mathbf{E}}{B^2}\cos(Bt) - \frac{M\mathbf{E}}{B^2} + \frac{M^2\mathbf{E}t}{B^2} - \frac{M^2\mathbf{E}}{B^2 B}\sin(Bt),$$

we find as the solution of the initial value problem of system (2)

$$\mathbf{v}(t) = \mathbf{E}t\left(1 + \frac{M^2}{B^2}\right) - \frac{M^2\mathbf{E}}{B^2 B}\sin(Bt) + \frac{M\mathbf{v}(0)}{B}\sin(Bt)$$

$$+ \frac{M\mathbf{E}}{B^2}(1 - \cos(Bt)) + \frac{M^2\mathbf{v}(0)}{B^2}(1 - \cos(Bt)) + \mathbf{v}(0)$$

where $\mathbf{v}(t = 0) \equiv \mathbf{v}(0)$.

Problem 3.   Solve the linear differential equation

$$\frac{d^2u}{dx^2} + u = 0 \tag{1}$$

with the following boundary conditions

$$u(0) = 1, \quad u(1) = 1, \tag{2a}$$

$$u(0) = 1, \quad u(\pi) = -1, \tag{2b}$$

$$u(0) = 1, \quad u(\pi) = -2. \tag{2c}$$

Discuss the result.

Solution.   The general solution to (1) is

$$u(x) = C_1\cos x + C_2\sin x$$

where C_1 and C_2 are the constants of integration. Imposing the boundary condition (2a) yields

$$u(0) = 1, u(1) = 1 \Rightarrow C_1 = 1, \quad C_2 = \frac{1 - \cos 1}{\sin 1}.$$

Thus with the boundary condition (2a) we have one solution. Imposing the boundary condition (2b) yields

$$u(0) = 1, \quad u(\pi) = -1 \Rightarrow C_1 = 1, \quad C_2 = \text{arbitrary}.$$

Since C_2 is arbitrary we have infinite solutions for boundary condition (2b). Imposing the boundary condition (2c) gives

$$u(0) = 1, \quad u(\pi) = -2 \Rightarrow C_1 = 1, \quad \underbrace{-2 = 1(-1) + C_2 \cdot 0}_{\text{cannot be satisfied}}.$$

Consequently, no solution exists for boundary condition (2c).

Remark.   *Boundary value problems can lead to* (a) *unique solutions,* (b) *arbitrarily many solutions,* (c) *no solution.*

Problem 4.   Let

$$H(\mathbf{p}, \mathbf{q}) = \frac{1}{2}\left(\frac{p_1^2}{m} + \frac{p_2^2}{m}\right) + \frac{1}{2}k(q_2 - q_1)^2 \tag{1}$$

be a Hamilton function, where k is a positive constant. Solve Hamilton's equations of motion for this Hamilton function.

Solution.   *Hamilton's equations of motion* are given by

$$\frac{dq_j}{dt} = \frac{\partial H}{\partial p_j}, \tag{2a}$$

$$\frac{dp_j}{dt} = -\frac{\partial H}{\partial q_j}, \tag{2b}$$

where $j = 1, 2$. Inserting the Hamilton function (1) into system (2) yields

$$\frac{dq_1}{dt} = \frac{p_1}{m}, \tag{3a}$$

$$\frac{dq_2}{dt} = \frac{p_2}{m}, \tag{3b}$$

$$\frac{dp_1}{dt} = k(-q_1 + q_2), \tag{3c}$$

$$\frac{dp_2}{dt} = k(q_1 - q_2). \tag{3d}$$

This is an autonomous system of first-order linear differential equations with constant coefficients. Let

$$A = \begin{pmatrix} 0 & 0 & \dfrac{1}{m} & 0 \\ 0 & 0 & 0 & \dfrac{1}{m} \\ -k & k & 0 & 0 \\ k & -k & 0 & 0 \end{pmatrix}.$$

Then the system (3) can be written in the matrix form

$$\begin{pmatrix} dq_1/dt \\ dq_2/dt \\ dp_1/dt \\ dp_2/dt \end{pmatrix} = A \begin{pmatrix} q_1 \\ q_2 \\ p_1 \\ p_2 \end{pmatrix}.$$

Thus the general solution of the initial value problem is given by

$$\begin{pmatrix} q_1(t) \\ q_2(t) \\ p_1(t) \\ p_2(t) \end{pmatrix} = e^{tA} \begin{pmatrix} q_1 \\ q_2 \\ p_1 \\ p_2 \end{pmatrix}\Bigg|_{q_1 \to q_1(0),\ldots,p_2 \to p_2(0)}.$$

The exponential function $\exp(tA)$ can be evaluated by determining the eigenvalues and normalized eigenvectors of A. The eigenvalues of A are given by

$$\lambda_{1,2} = 0, \quad \lambda_{3,4} = \pm i\omega$$

where

$$\omega^2 = \frac{2k}{m}.$$

Then the solution of the initial value problem is given by

$$q_1(t) = \frac{q_{10} - q_{20}}{2}\cos(\omega t) + \frac{p_{10} - p_{20}}{2\omega m}\sin(\omega t) + \frac{p_{10} + p_{20}}{2m}t + \frac{q_{10} + q_{20}}{2},$$

$$q_2(t) = \frac{q_{20} - q_{10}}{2}\cos(\omega t) - \frac{p_{10} - p_{20}}{2\omega m}\sin(\omega t) + \frac{p_{10} + p_{20}}{2m}t + \frac{q_{10} + q_{20}}{2}$$

where $q_{j0} \equiv q_j(t = 0)$ and $p_{j0} \equiv p_j(t = 0)$ are the initial values. The momenta p_1 and p_2 are given by (3a) and (3b), respectively.

Problem 5.   The *time-independent Schrödinger equation* (eigenvalue equation) for a one-particle problem is given by

$$\left(-\frac{\hbar^2}{2m}\Delta + U(\mathbf{r})\right) u = Eu \tag{1}$$

where

$$\Delta := \frac{\partial^2}{\partial x_1^2} + \frac{\partial^2}{\partial x_2^2} + \frac{\partial^2}{\partial x_3^2}.$$

Show that if the potential energy $U(\mathbf{r})$ can be written as a sum of functions of a single coordinate,

$$U(\mathbf{r}) = U_1(x_1) + U_2(x_2) + U_3(x_3)$$

then the time-independent Schrödinger equation can be decomposed into a set of one-dimensional equations of the form

$$\frac{d^2 u_j(x_j)}{dx_j^2} + \frac{2m}{\hbar^2}(E_j - U_j(x_j))u_j(x_j) = 0, \quad j = 1, 2, 3$$

with the help of the *separation ansatz*

$$u(\mathbf{r}) = u_1(x_1)u_2(x_2)u_3(x_3) \tag{2}$$

and

$$E = E_1 + E_2 + E_3.$$

Solution.   Substituting the ansatz (2) into (1) and dividing by $u_1 u_2 u_3$ we obtain

$$\sum_{j=1}^{3} \left(\frac{1}{u_j}\frac{d^2 u_j}{dx_j^2} - \frac{2m}{\hbar^2}U_j(x_j)\right) = -\frac{2m}{\hbar^2}E.$$

Since the terms in each bracket of the sum contain independent variables the equality can be valid for all (x_1, x_2, x_3) only if each bracket is a constant i.e. if

$$\frac{1}{u_j}\frac{d^2 u_j}{dx_j^2} - \frac{2m}{\hbar^2}U_j = -\frac{2m}{\hbar^2}E_j, \quad j = 1, 2, 3$$

where the E_j are constants, and $E = E_1 + E_2 + E_3$.

Problem 6. (i) Show that the general solution of the one-dimensional wave equation

$$\frac{1}{c^2}\frac{\partial^2 u}{\partial t^2} = \frac{\partial^2 u}{\partial x^2} \tag{1}$$

is given by

$$u(x,t) = f(x - ct) + g(x + ct) \tag{2}$$

where f and g are smooth functions and c is a positive constant.

(ii) Solve the initial value problem

$$u(t = 0, x) = u_0(x),$$

$$\frac{\partial u}{\partial t}(t = 0, x) = u_1(x).$$

Solution. (i) Let $s := x - ct$ and $r := x + ct$. Then

$$\frac{\partial u}{\partial t} = \frac{df}{ds}\frac{\partial s}{\partial t} + \frac{dg}{dr}\frac{\partial r}{\partial t} = -c\frac{df}{ds} + c\frac{dg}{dr}$$

and

$$\frac{\partial^2 u}{\partial t^2} = c^2\frac{d^2 f}{ds^2} + c^2\frac{d^2 g}{dr^2}. \tag{3}$$

Analogously,

$$\frac{\partial^2 u}{\partial x^2} = \frac{d^2 f}{ds^2} + \frac{d^2 g}{dr^2}. \tag{4}$$

Inserting (3) and (4) into (1) shows that (2) is the (general) solution. Equation (2) is the general solution of the one-dimensional wave equation since f and g are arbitrary.

(ii) From (2) we obtain

$$u_0(x) = f(x) + g(x).$$

Since

$$\frac{\partial u(x,t)}{\partial t} = -c\frac{df(x - ct)}{ds} + c\frac{dg(x + ct)}{dr},$$

follows that

$$u_1(x) = -c\frac{df(x)}{dx} + c\frac{dg(x)}{dx}. \tag{5}$$

Integrating (5) yields

$$A + \int_a^x u_1(\alpha)d\alpha = -cf(x) + cg(x)$$

where A, a are two arbitrary constants. Consequently

$$2cf(x) = cu_0(x) - A - \int_a^x u_1(\alpha)d\alpha , \tag{6a}$$

$$2cg(x) = cu_0(x) + A + \int_a^x u_1(\alpha)d\alpha . \tag{6b}$$

With the translation

$$x \to x - ct , \quad x \to x + ct$$

it follows from (6a) and (6b) that

$$2cf(x - ct) = cu_0(x - ct) - A - \int_a^{x-ct} u_1(\alpha)d\alpha , \tag{7a}$$

$$2cg(x + ct) = cu_0(x + ct) + A + \int_a^{x+ct} u_1(\alpha)d\alpha . \tag{7b}$$

Adding (7a) and (7b), we obtain

$$u(x, t) = \frac{1}{2}(u_0(x + ct) + u_0(x - ct)) + \frac{1}{2c} \int_{x-ct}^{x+ct} u_1(\alpha)d\alpha .$$

Problem 7.   The *three-dimensional wave equation* is given by

$$\frac{1}{c^2} \frac{\partial^2 u}{\partial t^2} = \frac{\partial^2 u}{\partial x^2} + \frac{\partial^2 u}{\partial y^2} + \frac{\partial^2 u}{\partial z^2} \equiv \Delta u . \tag{1}$$

(i) Express the wave equation in spherical coordinates. Omit the angle part and find the wave equation for the radial part.
(ii) Find the general solution of the wave equation which only includes the radial part.

Solution.   (i) The *spherical coordinates* are given by

$$x(r, \phi, \theta) = r \cos \phi \sin \theta , \quad y(r, \phi, \theta) = r \sin \phi \sin \theta , \quad z(r, \phi, \theta) = r \cos \theta$$

where $0 \le \phi < 2\pi$, $0 < \theta < \pi$ and $r > 0$. Let

$$u(x(r, \phi, \theta), y(r, \phi, \theta), z(r, \phi, \theta)) = v(r, \phi, \theta) .$$

Applying the chain rule we find from (1) that

$$\frac{1}{r^2}\frac{\partial}{\partial r}\left(r^2\frac{\partial}{\partial r}\right)v + \frac{1}{r^2\sin\theta}\frac{\partial}{\partial\theta}\left(\sin\theta\frac{\partial}{\partial\theta}\right)v + \frac{1}{r^2\sin^2\theta}\frac{\partial^2}{\partial\phi^2}v = \frac{1}{c^2}\frac{\partial^2 u}{\partial t^2}.$$

If v is a *spherically-symmetric solution* of the wave equation, i.e. $\partial v/\partial\theta = 0$ and $\partial v/\partial\phi = 0$ the wave equation becomes

$$\frac{1}{r^2}\frac{\partial}{\partial r}\left(r^2\frac{\partial v}{\partial r}\right) = \frac{1}{c^2}\frac{\partial^2 v}{\partial t^2}$$

or equivalently

$$\frac{1}{r^2}\left(2r\frac{\partial v}{\partial r} + r^2\frac{\partial^2 v}{\partial r^2}\right) = \frac{1}{c^2}\frac{\partial^2 v}{\partial t^2}.$$

Now we have the identity

$$\frac{\partial^2(rv)}{\partial r^2} \equiv \frac{\partial}{\partial r}\left(v + r\frac{\partial v}{\partial r}\right) \equiv \frac{\partial v}{\partial r} + \frac{\partial v}{\partial r} + r\frac{\partial^2 v}{\partial r^2} \equiv \frac{1}{r}\left(2r\frac{\partial v}{\partial r} + r^2\frac{\partial^2 v}{\partial r^2}\right).$$

Thus the wave equation takes the form

$$\frac{1}{r}\frac{\partial^2(rv)}{\partial r^2} = \frac{1}{c^2}\frac{\partial^2 v}{\partial t^2}$$

or

$$\frac{\partial^2(rv)}{\partial r^2} = \frac{1}{c^2}\frac{\partial^2(rv)}{\partial t^2}.$$

We define

$$\psi(r,t) := rv(r,t).$$

Then

$$\frac{\partial^2\psi}{\partial r^2} = \frac{1}{c^2}\frac{\partial^2\psi}{\partial t^2}.$$

In Problem 6 we have found the general solution, namely

$$\psi(r,t) = f_1(r - ct) + f_2(r + ct).$$

It follows that

$$v(r,t) = \frac{1}{r}f_1(r - ct) + \frac{1}{r}f_2(r + ct)$$

with $r > 0$. Let

$$v_0(r) = v(r,0), \quad v_1(r) = \frac{\partial v(r,0)}{\partial t}$$

be the initial distributions. Then

$$\psi_0(r) = \psi(r, 0) = (rv)(r, 0) = rv_0 \,,$$

$$\psi_1(r) = \frac{\partial \psi}{\partial t}(r, 0) = r\frac{\partial v}{\partial t}(r, 0) = rv_1(r) \,.$$

Using the method described in Problem 6 the solution to the initial value problem is given by

$$v(r, t) = \frac{1}{2r}(r + ct)v_0(r + ct) + \frac{1}{2r}(r - ct)v_0(r - ct) + \frac{1}{2cr}\int_{r-ct}^{r+ct} \alpha u_1(\alpha)d\alpha \,.$$

Problem 8.   Let

$$-\frac{\hbar}{i}\frac{\partial \psi}{\partial t} = \hat{H}\psi$$

be the *Schrödinger equation*, where

$$\hat{H} := -\frac{\hbar^2}{2m}\Delta + U(\mathbf{r}) \,,$$

$$\Delta := \frac{\partial^2}{\partial x_1^2} + \frac{\partial^2}{\partial x_2^2} + \frac{\partial^2}{\partial x_3^2} \,,$$

and $\mathbf{r} = (x_1, x_2, x_3)$. Let

$$\rho(\mathbf{r}, t) := \bar{\psi}(\mathbf{r}, t)\psi(\mathbf{r}, t) \tag{1}$$

where $\bar{\psi}$ denotes the complex conjugate of ψ. Find $\mathbf{j}$ such that

$$\mathrm{div}\mathbf{j} + \frac{\partial \rho}{\partial t} = 0 \tag{2}$$

where

$$\mathrm{div}\mathbf{j} := \frac{\partial j_1}{\partial x_1} + \frac{\partial j_2}{\partial x_2} + \frac{\partial j_3}{\partial x_3} \,.$$

Remark.   *Equation (2) is called a conservation law.*

Solution.   Since

$$-\frac{\hbar}{i}\frac{\partial \psi}{\partial t} = -\frac{\hbar^2}{2m}\Delta\psi + U\psi \,, \tag{3}$$

we obtain

$$\frac{\hbar}{i}\frac{\partial \bar{\psi}}{\partial t} = -\frac{\hbar^2}{2m}\Delta\bar{\psi} + U\bar{\psi} \,. \tag{4}$$

From (1) we obtain

$$\frac{\partial \rho}{\partial t} = \frac{\partial}{\partial t}(\bar{\psi}\psi) = \bar{\psi}\frac{\partial \psi}{\partial t} + \frac{\partial \bar{\psi}}{\partial t}\psi. \tag{5}$$

Inserting (3) and (4) into (5) gives

$$-\frac{\hbar}{i}\frac{\partial \rho}{\partial t} = \bar{\psi}\hat{H}\psi - \psi\hat{H}\bar{\psi}.$$

Now

$$\bar{\psi}\hat{H}\psi - \psi\hat{H}\bar{\psi} = -\frac{\hbar^2}{2m}(\bar{\psi}\Delta\psi - \psi\Delta\bar{\psi}) = -\frac{\hbar^2}{2m}\text{div}(\bar{\psi}\nabla\psi - \psi\nabla\bar{\psi}).$$

Here ∇ denotes the gradient. Thus

$$\mathbf{j} = \frac{\hbar}{2mi}(\bar{\psi}\nabla\psi - \psi\nabla\bar{\psi}).$$

Problem 9. (i) Show that the substitution

$$u = -\frac{1}{v}\frac{dv}{dt} \tag{1}$$

reduce the nonlinear differential equation (*Riccati equation*)

$$\frac{du}{dt} = u^2 + t, \quad u(0) = 1 \tag{2}$$

to the linear second-order differential equation

$$\frac{d^2v}{dt^2} + tv = 0. \tag{3}$$

(ii) Find the inverse transformation of (1).

Solution. (i) From (1) we obtain

$$\frac{du}{dt} = \frac{1}{v^2}\left(\frac{dv}{dt}\right)^2 - \frac{1}{v}\frac{d^2v}{dt^2}. \tag{4}$$

Inserting (4) and (1) into (2) yields (3).

(ii) Since

$$u(t) = -\frac{d}{dt}(\ln v),$$

we find

$$v(t) = \exp\left(-\int^{t} u(s)ds\right).$$

Equation (2) has no solution in terms of the elementary functions. *Bessel functions* are needed to solve it.

Chapter 13

Integration

Problem 1. Calculate

$$I = \int_0^\infty e^{-x^2} dx.$$ (1)

Solution. To evaluate the integral we transform the single integral into double integral. Thus

$$I^2 = \int_0^\infty \left[\int_0^\infty e^{-x^2} dx \right] e^{-y^2} dy$$

$$= \int_0^\infty \int_0^\infty e^{-x^2} e^{-y^2} dx dy = \int_0^\infty \int_0^\infty e^{-(x^2+y^2)} dx dy.$$ (2)

Introducing *polar coordinates*

$$x(r, \phi) = r \cos \phi, \quad y(r, \phi) = r \sin \phi$$ (3)

with $0 \le \phi < \pi/2$, $0 \le r < \infty$ we obtain

$$I^2 = \int_0^{\pi/2} \int_0^\infty e^{-r^2} r \, dr d\phi = \int_0^{\pi/2} \left. -\frac{1}{2} e^{-r^2} \right|_0^\infty d\phi = \frac{1}{2} \int_0^{\pi/2} d\phi = \frac{1}{4}\pi.$$ (4)

It follows that $I = \sqrt{\pi}/2$.

Problem 2. Given that

$$\int_0^\infty \frac{\sin x}{x} dx = \frac{1}{2}\pi.$$

Find

$$I = \int_0^\infty \frac{\sin^2 x}{x^2} dx .$$

Solution. We calculate the more general integral

$$I(\epsilon) = \int_0^\infty \frac{\sin^2(\epsilon x)}{x^2} dx , \quad \epsilon \geq 0$$

by using a technique called *parameter differentiation*. Differentiating each side of the previous equation with respect to ϵ (we can interchange differentiation and integration), we obtain

$$\frac{dI(\epsilon)}{d\epsilon} = \int_0^\infty \frac{2x \sin(\epsilon x) \cos(\epsilon x)}{x^2} dx = \int_0^\infty \frac{\sin(2\epsilon x)}{x} dx .$$

We set $y = 2\epsilon x$. Therefore we find $dy = 2\epsilon dx$, and

$$\frac{dI(\epsilon)}{d\epsilon} = \int_0^\infty \frac{\sin y}{y} dy = \frac{1}{2}\pi .$$

Integrating each side gives $I(\epsilon) = \frac{1}{2}\pi\epsilon + C$, where C is the constant of integration. Since $I(0) = 0$, we obtain $C = 0$. Thus

$$I(\epsilon) = \frac{1}{2}\pi\epsilon , \quad \epsilon \geq 0 .$$

Setting $\epsilon = 1$ yields $I(1) = I = \pi/2$.

Problem 3.   (i) Let $\alpha_j \in \mathbf{C}$. Calculate

$$I = \int_0^\infty d\tau_2 \int_0^{\tau_2} d\tau_1 \exp(\alpha_1 \tau_1 + \alpha_2 \tau_2) \tag{1}$$

where $\Re\alpha_2 < 0$ and $\Re(\alpha_1 + \alpha_2) < 0$. Here $\Re$ denotes the real part.

(ii) Calculate

$$I = \int_0^\infty d\tau_n \int_0^{\tau_n} d\tau_{n-1} \cdots \int_0^{\tau_2} d\tau_1 \exp\left(\sum_{m=1}^n \alpha_m \tau_m\right) \tag{2}$$

where

$$\Re \sum_{m=k}^n \alpha_m < 0$$

for all $k \leq n$.

Solution. (i) Equation (1) can be written in the form

$$\int_0^\infty d\tau_2 \int_0^{\tau_2} d\tau_1 \exp(\alpha_1\tau_1 + \alpha_2\tau_2)$$

$$= \int_0^\infty d\tau_2 \int_0^\infty d\tau_1 \Theta(\tau_2 - \tau_1) \exp(\alpha_1\tau_1 + \alpha_2\tau_2)$$

where

$$\Theta(\tau_2 - \tau_1) = \begin{cases} 1 & \text{for } \tau_2 > \tau_1, \\ 0 & \text{otherwise}. \end{cases}$$

Θ is called the *step function*. By the substitution

$$x_1 = \tau_1, \quad x_2 = \tau_2 - \tau_1,$$

we obtain

$$I = \left(\int_0^\infty dx_2 e^{\alpha_2 x_2} \right) \left(\int_0^\infty dx_1 e^{(\alpha_1 + \alpha_2)x_1} \right) = \frac{1}{\alpha_2(\alpha_1 + \alpha_2)}.$$

(ii) By the substitution

$$x_{n-1} = \tau_{n-1}, \quad x_n = \tau_n - \tau_{n-1}$$

we write (2) in the form

$$I = \left(\int_0^\infty dx_n \exp(\alpha_n x_n) \right) \left(\int_0^\infty dx_{n-1} \int_0^{x_{n-1}} d\tau_{n-2} \cdots \int_0^{\tau_2} d\tau_1 \right.$$

$$\left. \times \exp\left(\sum_{m=1}^{n-2} \alpha_m \tau_m \right) \exp((\alpha_n + \alpha_{n-1})x_{n-1}) \right).$$

Repeating this process we can write the integral

$$I = \left(\int_0^\infty dx_n \exp(\alpha_n x_n) \right) \left(\int_0^\infty dx_{n-1} \exp(\alpha_n + \alpha_{n-1})x_{n-1} \right) \cdots$$

$$\times \left(\int_0^\infty dx_1 \exp\left(\left(\sum_{m=1}^n \alpha_m \right) x_1 \right) \right).$$

We obtain

$$I = (-1)^n (\alpha_n)^{-1}(\alpha_n + \alpha_{n-1})^{-1} \cdots \left(\sum_{m=1}^n \alpha_m \right)^{-1}$$

since for $s < 0$ we have

$$\int_0^\infty dx e^{sx} = \frac{1}{s} e^{sx} \Big|_0^\infty = -\frac{1}{s}.$$

Problem 4.   Let A be an $n \times n$ Hermitian matrix. Prove the identity

$$e^{A^2} \equiv \int_{-\infty}^{+\infty} dx \exp(-\pi x^2 I - 2Ax\sqrt{\pi}) \tag{1}$$

where I is the $n \times n$ unit matrix.

Solution.   Since A is an $n \times n$ hermitian matrix there is an $n \times n$ unitary matrix U such that the matrix U^*AU is diagonal, where $U^* = U^{-1}$. We write

$$\tilde{A} = U^*AU = \text{diag}(\lambda_1, \lambda_2, \ldots, \lambda_n). \tag{2}$$

Obviously, $\lambda_1, \ldots, \lambda_n$ are the eigenvalues of the matrix A. Since A is hermitian the eigenvalues are real. From (1) we obtain

$$U^* e^{A^2} U \equiv U^* \left(\int_{-\infty}^{+\infty} dx \exp(-\pi x^2 I - 2Ax\sqrt{\pi}) \right) U. \tag{3}$$

Since

$$U^* e^{A^2} U \equiv e^{U^* A^2 U} \equiv e^{U^* AUU^* AU} \equiv e^{\tilde{A}^2}$$

and

$$U^* \left(\int_{-\infty}^{+\infty} dx \exp(-\pi x^2 I - 2Ax\sqrt{\pi}) \right) U = \int_{-\infty}^{+\infty} dx \exp(-\pi x^2 I - 2\tilde{A}x\sqrt{\pi}),$$

we obtain from identity (3) that

$$e^{\tilde{A}^2} \equiv \int_{-\infty}^{+\infty} dx \exp(\pi x^2 I - 2\tilde{A}x\sqrt{\pi}). \tag{4}$$

From (2) we obtain

$$\tilde{A}^2 = \text{diag}(\lambda_1^2, \lambda_2^2, \ldots, \lambda_n^2)$$

and

$$e^{\tilde{A}^2} = \text{diag}(e^{\lambda_1^2}, e^{\lambda_2^2}, \ldots, e^{\lambda_n^2}).$$

For the right-hand side of (4) we find

$$\int_{-\infty}^{+\infty} dx \, \exp(-\pi x^2 I - 2\tilde{A}x\sqrt{\pi})$$

$$= \int_{-\infty}^{+\infty} dx \, \exp\left(\operatorname{diag}(-\pi x^2 - 2\lambda_1 x\sqrt{\pi}, \right.$$

$$\left. -\pi x^2 - 2\lambda_2 x\sqrt{\pi}, \ldots, -\pi x^2 - 2\lambda_n x\sqrt{\pi})\right)$$

$$= \int_{-\infty}^{+\infty} dx \, \operatorname{diag}(e^{-\pi x^2 - 2\lambda_1 x\sqrt{\pi}}, e^{-\pi x^2 - 2\lambda_2 x\sqrt{\pi}}, \ldots, e^{-\pi x^2 - 2\lambda_n x\sqrt{\pi}})$$

$$= \operatorname{diag}\left(\int_{-\infty}^{+\infty} e^{-\pi x^2 - 2\lambda_1 x\sqrt{\pi}} dx, \int_{-\infty}^{+\infty} e^{-\pi x^2 - 2\lambda_2 x\sqrt{\pi}} dx, \ldots, \right.$$

$$\left. \times \int_{-\infty}^{+\infty} e^{-\pi x^2 - 2\lambda_n x\sqrt{\pi}} dx\right).$$

Since

$$\int_{-\infty}^{+\infty} e^{-\pi x^2 - 2\lambda_i x\sqrt{\pi}} dx = \exp(\lambda_i^2),$$

we have

$$\int_{-\infty}^{+\infty} dx \, \exp(-\pi x^2 I - 2\tilde{A}x\sqrt{\pi}) \equiv \operatorname{diag}(e^{\lambda_1^2}, e^{\lambda_2^2}, \ldots, e^{\lambda_n^2}).$$

This proves the identity (3). Since $U^* = U^{-1}$ we also proved identity (1).

Problem 5.   Given the nonlinear system of differential equations

$$\frac{d^2\theta}{ds^2} - \frac{2b\sin\phi}{a + b\cos\phi}\frac{d\theta}{ds}\frac{d\phi}{ds} = 0, \tag{1a}$$

$$\frac{d^2\phi}{ds^2} + \frac{(a + b\cos\phi)\sin\phi}{b}\left(\frac{d\theta}{ds}\right)^2 = 0, \tag{1b}$$

where $a > b > 0$. Show that

$$\frac{d^2\phi}{ds^2} + \frac{C^2\sin\phi}{b(a + b\cos\phi)^3} = 0 \tag{2}$$

where C is a constant of integration.

Solution.   From (1) we find

$$\frac{\frac{d^2\theta}{ds^2}}{\frac{d\theta}{ds}} = \frac{2b\sin\phi}{a+b\cos\phi}\frac{d\phi}{ds}$$

or

$$\frac{d}{ds}\left(\ell n\frac{d\theta}{ds}\right) = \frac{2b\sin\phi}{a+b\cos\phi}\frac{d\phi}{ds}.$$

Therefore

$$\int\frac{d}{ds}\left(\ell n\frac{d\theta}{ds}\right)ds = \int\frac{2b\sin\phi}{a+b\cos\phi}\frac{d\phi}{ds}ds = \int\frac{2b\sin\phi}{a+b\cos\phi}d\phi$$

or

$$\ell n\frac{d\theta}{ds} = -2\ell n|a+b\cos\phi| + K \tag{3}$$

where K is the constant of integration. From (3) we obtain

$$\frac{d\theta}{ds} = \frac{C}{(a+b\cos\phi)^2} \tag{4}$$

where $C = e^K$. Inserting (4) into (1b) yields (2).

Problem 6.   Calculate

$$I(\lambda) = \int_{|\mathbf{q}|<k_F}\frac{dq_1\,dq_2\,dq_3}{|\mathbf{k}-\mathbf{q}|^2 + \lambda^2} \tag{1}$$

where

$$|\mathbf{k}-\mathbf{q}|^2 = (k_1 - q_1)^2 + (k_2 - q_2)^2 + (k_3 - q_3)^2$$

and $\lambda^2 > 0$.

Solution.   Introducing spherical coordinates (q, ϕ, θ) we obtain

$$I(\lambda) = \int_0^{k_F} q^2\,dq\int_0^{\pi}\sin\theta\,d\theta\int_0^{2\pi}d\phi\frac{1}{k^2 - 2qk\cos\theta + q^2 + \lambda^2}$$

where θ is given by

$$\mathbf{k}\cdot\mathbf{q} = kq\cos\theta.$$

We set $u = \cos\theta$. Then $du = -\sin\theta d\theta$ and $\theta = 0 \to u = 1$, $\theta = \pi \to u- =$ 1. We obtain

$$I(\lambda) = -2\pi \int_0^{k_F} q^2 dq \int_1^{-1} du \frac{1}{k^2 + q^2 - 2qku + \lambda^2} \cdot$$

The integration with respect to u can easily be performed. We find

$$I(\lambda) = 2\pi \int_0^{k_F} q^2 dq \left[\frac{1}{-2qk} \ln(k^2 + q^2 + \lambda^2 - 2qku) \right]_{u=1}^{u=-1}.$$

Thus

$$I(\lambda) = -\frac{\pi}{k} \int_0^{k_F} q(\ln(k^2 + q^2 + \lambda^2 + 2qk) - \ln(k^2 + q^2 + \lambda^2 - 2qk))dq$$

and

$$I(\lambda) = -\frac{\pi}{k} \underbrace{\int_0^{k_F} q \ln(\lambda^2 + (k+q)^2)dq}_{F_1} + \frac{\pi}{k} \underbrace{\int_0^{k_F} q \ln(\lambda^2 + (k-q)^2)dq}_{F_2}.$$

Now we have to calculate the integrals F_1 and F_2. We find

$$F_1 = \int_{q=0}^{k_F} q \ln(\lambda^2 + (k+q)^2)dq = \int_{x=k}^{k_F+k} (x-k) \ln(\lambda^2 + x^2)dx$$

or

$$F_1 = -\underbrace{\int_{x=k}^{k_F+k} k \ln(\lambda^2 + x^2)dx}_{F_1'} + \underbrace{\int_{x=k}^{k_F+k} x \ln(\lambda^2 + x^2)dx}_{F_1''}.$$

Now we have to calculate F_1' and F_1''. Since

$$\int \ln(\lambda^2 + x^2)dx = x \ln(x^2 + \lambda^2) + 2\lambda \arctan\frac{x}{\lambda} - 2x,$$

we find for F_1'

$$F_1' = -k[(k_F + k) \ln((k_F + k)^2 + \lambda^2) + 2\lambda \arctan\left(\frac{k_F + k}{\lambda}\right) - 2(k_F + k)$$

$$- k \ln(k^2 + \lambda^2) - 2\lambda \arctan\left(\frac{k}{\lambda}\right) + 2k].$$

Since

$$\int x f(x^2) dx = \frac{1}{2} \int f(u) du$$

with $x^2 = u$ we find

$$F_1'' = \int_{x=k}^{k_F+k} x \ln(\lambda^2 + x^2) dx = \frac{1}{2} \int_{u=k^2}^{(k_F+k)^2} \ln(\lambda^2 + u) du.$$

Thus

$$F_1'' = \frac{1}{2}(((k_F+k)^2 + \lambda^2) \ln((k_F+k)^2 + \lambda^2) - (k_F+k)^2$$
$$- (k^2 + \lambda^2) \ln(k^2 + \lambda^2) + k^2).$$

The integral F_2 is given by

$$F_2 = \int_{q=0}^{k_F} q \ln(\lambda^2 + (q-k)^2) dq.$$

It follows that

$$F_2 = \int_{x=-k}^{k_F-k} (x+k) \ln(\lambda^2 + x^2) dx$$

$$= \underbrace{\int_{x=-k}^{k_F-k} k \ln(\lambda^2 + x^2) dx}_{F_2'} + \underbrace{\int_{x=-k}^{k_F-k} x \ln(\lambda^2 + x^2) dx}_{F_2''}.$$

Now we have to calculate F_2'. We obtain

$$F_2' = \int_{x=-k}^{k_F-k} k \ln(\lambda^2 + x^2) dx.$$

Thus

$$F_2' = k(k_F - k) \ln((k_F - k)^2 + \lambda^2) + 2k\lambda \arctan\left(\frac{k_F - k}{\lambda}\right)$$

$$- 2k(k_F - k) + k^2 \ln(k^2 + \lambda^2) - 2k\lambda \arctan\left(\frac{-k}{\lambda}\right) - 2k^2.$$

For F_2'' we obtain

$$F_2'' = \int_{x=-k}^{k_F-k} x \ln(\lambda^2 + x^2) dx = \frac{1}{2} \int_{u=k^2}^{(k_F-k)^2} \ln(\lambda^2 + u) du$$

or

$$F_2'' = \frac{1}{2}(((k_F - k)^2 + \lambda^2)\ln((k_F - k)^2 + \lambda^2) - (k_F - k)^2$$
$$- (k^2 + \lambda^2)\ln(k^2 + \lambda^2) + k^2)$$

where we have used $u = x^2$ and therefore $du = 2xdx$.

Remark. *The integral* (1) *plays an important role in solid state physics. Here $\hbar k_f$ is the* Fermi momentum. *The absolute value of momentum of particles at zero temperature is called the Fermi momentum and the region of momentum space with momentum $\hbar k_f$ is called the Fermi surface.*

Problem 7. Calculate the integral (which plays a rôle in electrostatic fields)

$$U(r) = \frac{1}{4\pi} \int_{r'} \int_{\theta=0}^{\pi} \int_{\phi=0}^{2\pi} \frac{\rho(r')}{(r^2 + r'^2 - 2rr'\cos\theta)^{1/2}} r'^2 \sin\theta dr' d\theta d\phi .$$

Solution. The integration with respect to ϕ can easily be performed. We find

$$U(r) = \frac{1}{2} \int_{r'} \int_{\theta=0}^{\pi} \frac{\rho(r')}{(r^2 + r'^2 - 2rr'\cos\theta)^{1/2}} r'^2 \sin\theta dr' d\theta .$$

To perform the θ-integration we set $u = \cos\theta$. Therefore, $du = -\sin\theta d\theta$ and

$$U(r) = -\frac{1}{2} \int_{r'} \int_{u=1}^{-1} \frac{\rho(r')}{(r^2 + r'^2 - 2rr'u)^{1/2}} r'^2 dr' du ,$$

$$U(r) = \frac{1}{2} \int_{r'} \rho(r') \left(\frac{(r^2 + r'^2 + 2rr')^{1/2}}{rr'} - \frac{(r^2 + r'^2 - 2rr')^{1/2}}{rr'} \right) r'^2 dr' .$$

Since

$$(r^2 + r'^2 - 2rr')^{1/2} = ((r - r')^2)^{1/2} = r - r' \quad \text{for } r \geq r' ,$$
$$(r^2 + r'^2 - 2rr')^{1/2} = ((r' - r)^2)^{1/2} = r' - r \quad \text{for } r' \geq r ,$$

we obtain

$$U(r) = \frac{1}{2} \int_{r'=0}^{r} \rho(r') \frac{(r + r') - (r - r')}{r} r' dr'$$
$$+ \frac{1}{2} \int_{r'=r}^{\infty} \rho(r') \frac{(r + r') - (r' - r)}{r} r' dr' .$$

Consequently,

$$U(r) = \frac{1}{r} \int_{r'=0}^{r} \rho(r')r'^2 dr' + \int_{r'=r}^{\infty} \rho(r')r' dr'.$$

Problem 8.   Evaluate the *Lebesgue integral*

$$\int_0^1 x^2 dx.$$ (1)

Solution.   Let E be the set $E = [0,1)$. Then E is an elementary set having Lebesgue measure 1. The function $f : \mathbf{R} \to \mathbf{R}$ defined by

$$f(x) = \begin{cases} x^2 & \text{if } x \in E, \\ 0 & \text{if } x \notin E, \end{cases}$$ (2)

is a measurable function since

$$\{x : f(x) \geq a\}$$ (3)

is measurable for all $a \in \mathbf{R}$. We introduce a monotonic increasing sequence of simple functions tending to f as follows. Let

$$Q_{p,s} := \left\{ x : \frac{p-1}{2^s} \leq x < \frac{p}{2^s} \right\}$$ (4)

where

$$p = 1, 2, 4, \ldots, 2^s, \quad s = 1, 2, \ldots.$$ (5)

We define

$$f_s(x) := \begin{cases} \left(\dfrac{p-1}{2^s}\right)^2 & \text{for } x \in Q_{p,s}, \\ 0 & \text{for } x \in \mathbf{R} \setminus E. \end{cases}$$ (6)

Then for all $x \in \mathbf{R}$ we have

$$0 \leq f_s(x) \leq f(x).$$ (7)

Moreover, for all $x \in \mathbf{R}$, we find

$$f_{s+1}(x) \geq f_s(x).$$ (8)

This means f_s is a monotonic increasing function with increasing s. Furthermore

$$0 \le f(x) - f_s(x) \le \left(\frac{2^s}{2^s}\right)^2 - \left(\frac{2^s - 1}{2^s}\right)^2 = \frac{2^{2s} - 1}{2^s 2^s} < \frac{2}{2^s}$$

so that

$$f_s(x) \to f(x) \quad \text{as } s \to \infty.$$

Therefore

$$\int_E f_s(x)dx = \frac{1}{2^s}\left(0 + \left(\frac{1}{2^s}\right)^2 + \left(\frac{2}{2^s}\right)^2 + \cdots + \left(\frac{2^s - 1}{2^s}\right)^2\right).$$

Thus

$$\int_E f_s(x)dx = \frac{1}{n^3}(1 + 2^2 + 3^2 + \cdots + (n-1)^2) = \frac{1}{n^3}\left(\frac{(n-1)n(2n-1)}{6}\right)$$

where we used $n = 2^s$. Therefore

$$\int_E f_s(x)dx \to \frac{1}{3} \quad \text{as } s \to \infty.$$

It follows that

$$\int_0^1 x^2 dx = \frac{1}{3}.$$

Problem 9.   Two friends plan to meet at the library during a given 1-hour period. Their arrival times are independent and randomly distributed across the 1-hour period. Each agrees to wait for 15 minutes, or until the end of the hour. If the friend has not appeared during that time, the other friend will leave. Find the probability that the friends will meet.

Solution.   If X_1 denotes one person's arrival time in $[0, 1]$, the 1-hour period, and X_2 denotes the second person's arrival time, then (X_1, X_2) can be modeled as having a two-dimensional uniform distribution over the unit square. That is,

$$f(x_1, x_2) = \begin{cases} 1 & 0 \le x_1 \le 1, 0 \le x_2 \le 1, \\ 0 & \text{otherwise}. \end{cases}$$

The event that the two friends will meet depends upon the time, U, between their arrivals, where

$$U = |X_1 - X_2|.$$

Next we find the distribution function for the random variable U. For

$$U \leq u,$$

we have

$$|X_1 - X_2| \leq u.$$

Therefore

$$-u \leq X_1 - X_2 \leq u.$$

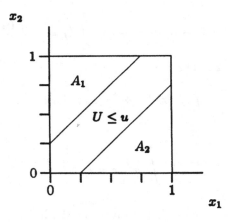

The figure shows that the region over which (X_1, X_2) has positive probability is the region defined by $U \leq u$. The probability that $U \leq u$ can be found by integrating f over the six-sided region shown in the figure. This can be simplified by integrating over the triangles (A_1 and A_2) and subtracting from one. We have

$$F_U(u) = P(U \leq u) = \int\!\!\int_{|x_1 - x_2| \leq u} f(x_1, x_2) dx_1 dx_2.$$

Thus

$$F_U(u) = 1 - \int\!\!\int_{A_1} f(x_1, x_2) dx_1 dx_2 - \int\!\!\int_{A_2} f(x_1, x_2) dx_1 dx_2.$$

Therefore

$$F_U(u) = 1 - \int_u^1 \int_0^{x_2-u} dx_1 dx_2 - \int_0^{1-u} \int_{x_2+u}^1 dx_1 dx_2 .$$

Finally, we arrive at

$$F_U(u) = 1 - (1-u)^2, \quad 0 \le u \le 1.$$

Thus

$$P\left(U \le \frac{15}{60}\right) = 1 - \left(1 - \frac{15}{60}\right)^2 = \frac{7}{16} .$$

Problem 10.   Let $a, b \in \mathbf{R}$, $a, b \ne 0$ and $b > a$. Show that

$$\frac{1}{ab} = \int_0^1 \frac{dz}{(az + b(1-z))^2} . \tag{1}$$

If a and b have opposite sign, z is to be considered as a complex variable and the path of the integration must deviate from the real axis so as to avoid the singularity at $z^* = b/(b-a)$.

Solution.   Identity (1) is obtained by noting that

$$\frac{1}{ab} = \frac{1}{b-a}\left(\frac{1}{a} - \frac{1}{b}\right) = \frac{1}{b-a}\int_a^b \frac{dx}{x^2}$$

and introducing in the last integral the new variable z via

$$x = az + b(1-z).$$

Thus

$$dx = (a-b)dz$$

and

$$z = 0 \Rightarrow x = b, \quad z = 1 \Rightarrow x = a.$$

The identity (1) holds for all values of a and b with $a, b \ne 0$. If, however, a and b are of the opposite sign, z is to be considered as a complex variable and the path of integration must deviate from the real axis so as to avoid the singularity (pole) at

$$z^* = \frac{b}{b-a} .$$

Any path joining $z = 0$ and $z = 1$ but not passing through the singularity may actually be chosen since the residue of the integrand at the singularity vanishes.

Hint. The integral given by identity (1) is a special case of the following general identity

$$\frac{1}{a_1 a_2 \cdots a_n} = (n-1)! \int_0^1 \cdots \int_0^1 \frac{dz_1 dz_2 \cdots dz_n}{(a_1 z_1 + a_2 z_2 + \cdots + a_n z_n)^n}$$

where

$$\sum_{i=1}^n z_i = 1.$$

To evaluate this integral we can use

$$\frac{1}{(n-1)!} \frac{1}{a_1 a_2 \cdots a_n}$$

$$= \int_0^1 dz_1 \int_0^{z_1} dz_2 \cdots \int_0^{z_{n-2}} dz_{n-1}$$

$$\times \frac{1}{(a_n z_{n-1} + a_{n-1}(z_{n-2} - z_{n-1}) + \cdots + a_1(1 - z_1))^n}$$

$$= \int_0^1 \epsilon_1^{n-2} d\epsilon_1 \int_0^1 \epsilon_2^{n-3} d\epsilon_2 \cdots \int_0^1 d\epsilon_{n-1}$$

$$\times \frac{1}{(a_1 \epsilon_1 \epsilon_2 \cdots \epsilon_{n-1} + a_2 \epsilon_1 \cdots \epsilon_{n-2}(1 - \epsilon_{n-1}) + \cdots + a_n(1 - \epsilon_1))^n} \, .$$

Chapter 14

Continuous Fourier Transform

Problem 1. Let $a > 0$. We define

$$f_a(x) = \begin{cases} \dfrac{1}{2a} & \text{for } |x| < a, \\ 0 & \text{for } |x| > a. \end{cases}$$

i) Calculate

$$\int_{\mathbf{R}} f_a(x)dx.$$

ii) Calculate the *Fourier transform* of f_a. The Fourier transform $\hat{f}_a(k)$ of $f_a(x)$ is defined as

$$\hat{f}_a(k) := \int_{\mathbf{R}} f_a(x)e^{ikx}dx.$$

Discuss the cases: a large and a small.

Remark. *The inverse transform is given by*

$$f_a(x) = \frac{1}{2\pi} \int_{\mathbf{R}} \hat{f}_a(k)e^{-ikx}dk.$$

Solution. (i) We find

$$\int_{\mathbf{R}} f_a(x)dx = \frac{1}{2a} \int_{-a}^{a} dx = 1.$$

Thus the integral is independent of a.

141

(ii) From (3) we obtain

$$\hat{f}_a(k) = \int_{-a}^{a} \frac{e^{ikx}}{2a} dx = \frac{1}{2a} \int_{-a}^{a} e^{ikx} dx.$$

Thus

$$\hat{f}_a(k) = \frac{1}{2aik} \left| e^{ikx} \right|_{-a}^{a}.$$

Finally

$$\hat{f}_a(k) = \frac{1}{ak} \frac{e^{ika} - e^{-ika}}{2i} = \frac{\sin(ak)}{ak}.$$

For $k = 0$ we find

$$\hat{f}_a(0) = 1.$$

Problem 2.   Let $N \in \mathbf{N}$ and let

$$V(t) = \begin{cases} V_0 e^{i\omega_0 t} & \text{if } nT \le t \le (nT + \tau) \text{ for } n = 0, 1, 2, \ldots, N-1, \\ 0 & \text{otherwise} \end{cases}$$

where V_0 and τ are positive constants. Calculate the Fourier transform.

Solution.   For general N the Fourier transform is given by

$$S(\omega) = V_0 \sum_{n=0}^{N-1} \int_{nT}^{nT+\tau} e^{-i\omega_0 t} e^{i\omega t} dt.$$

Thus

$$S(\omega) = V_0 \frac{e^{i(\omega-\omega_0)\tau}}{i(\omega - \omega_0)} \sum_{n=0}^{N-1} e^{in(\omega-\omega_0)T}.$$

The sum over n is a geometric series, which can be evaluated as follows

$$\sum_{n=0}^{N-1} e^{in\alpha} = \frac{1 - e^{iN\alpha}}{1 - e^{i\alpha}} = e^{i(N-1)\alpha/2} \frac{\sin(N\alpha/2)}{\sin(\alpha/2)}.$$

With

$$\alpha := (\omega - \omega_0)T,$$

we obtain

$$S(\omega) = V_0 \frac{e^{i(\omega-\omega_0)\tau}}{i(\omega - \omega_0)} e^{i(N-1)(\omega-\omega_0)/2} \frac{\sin(N(\omega - \omega_0)T/2)}{\sin((\omega - \omega_0)T/2)}.$$

Problem 3. Calculate the Fourier transform of

$$f(\mathbf{r}) = \frac{1}{r^2 + \lambda^2}$$

where $\mathbf{r} = (x, y, z)$,

$$\mathbf{r}^2 \equiv r^2 = x^2 + y^2 + z^2$$

and $\lambda^2 > 0$.

Solution. The Fourier transform is defined as

$$\hat{f}(\mathbf{k}) := \int_{-\infty}^{+\infty} \int_{-\infty}^{+\infty} \int_{-\infty}^{+\infty} \frac{e^{i\mathbf{k}\cdot\mathbf{r}}}{r^2 + \lambda^2} dx\,dy\,dz \tag{1}$$

where $\mathbf{k} \cdot \mathbf{r} = k_1 x + k_2 y + k_3 z$. Introducing spherical coordinates

$$x(r, \phi, \theta) = r \cos\phi \sin\theta\,,$$
$$y(r, \phi, \theta) = r \sin\phi \sin\theta\,,$$
$$z(r, \phi, \theta) = r \cos\theta\,,$$

we obtain

$$dx\,dy\,dz = r^2 \sin\theta\,dr\,d\theta\,d\phi$$

where $0 \le \phi < 2\pi$, $0 \le \theta < \pi$, and $r \ge 0$. We can set

$$\mathbf{k} \cdot \mathbf{r} = kr \cos\theta$$

where $k = \|\mathbf{k}\|$. Now (1) takes the form

$$\hat{f}(\mathbf{k}) = \int_{r=0}^{\infty} \int_{\theta=0}^{\pi} \int_{\phi=0}^{2\pi} \frac{e^{ikr\cos\theta}}{r^2 + \lambda^2} r^2 \sin\theta\,dr\,d\theta\,d\phi\,.$$

The integration over ϕ can be easily performed and we find

$$\hat{f}(\mathbf{k}) = 2\pi \int_{r=0}^{\infty} \int_{\theta=0}^{\pi} \frac{e^{ikr\cos\theta}}{r^2 + \lambda^2} r^2 \sin\theta\,dr\,d\theta\,. \tag{2}$$

We set

$$u = \cos\theta$$

and therefore

$$du = -\sin\theta\,d\theta\,.$$

It follows that $\theta = 0 \rightarrow u = 1$ and $\theta = \pi \rightarrow u = -1$. Then (2) takes the form

$$\hat{f}(\mathbf{k}) = -2\pi \int_{r=0}^{\infty} \frac{r^2}{r^2 + \lambda^2} \left(\int_{u=1}^{u=-1} e^{ikru} du \right) dr .$$

The integration over u can easily be performed. We find

$$\int_{1}^{-1} e^{ikru} du = \frac{-2\sin(kr)}{kr} .$$

Thus we arrive at

$$\hat{f}(\mathbf{k}) = \frac{4\pi}{k} \int_{r=0}^{\infty} \frac{r\sin(kr)}{r^2 + \lambda^2} dr .$$

The integral on the right-hand side can be solved by applying the integration in the complex plane. We obtain

$$\hat{f}(\mathbf{k}) = \frac{2\pi^2}{k} e^{-\lambda k} .$$

Problem 4.　Let $f, g \in L_2(\mathbf{R})$, where $L_2(\mathbf{R})$ denotes the Hilbert space of the square-integrable functions over $\mathbf{R}$. Let

$$\hat{f}(k) = \int_{\mathbf{R}} f(x) e^{ikx} dx$$

and

$$\hat{g}(k) = \int_{\mathbf{R}} g(x) e^{ikx} dx .$$

Show that

$$(f, g) = \frac{1}{2\pi} (\hat{f}, \hat{g}) \tag{1}$$

where $(,)$ denotes the scalar product in the Hilbert space $L_2(\mathbf{R})$.

Solution.　We have

$$(f, g) := \int_{\mathbf{R}} \bar{f}(x) g(x) dx = \frac{1}{2\pi} \int_{\mathbf{R}} \bar{f}(x) \left(\int_{\mathbf{R}} \hat{g}(k) e^{-ikx} dk \right) dx .$$

Thus

$$(f, g) = \frac{1}{2\pi} \int_{\mathbf{R}} \hat{g}(k) \left(\int_{\mathbf{R}} \bar{f}(x) e^{-ikx} dx \right) dk = \frac{1}{2\pi} \int_{\mathbf{R}} \bar{\hat{f}}(k) \hat{g}(k) dk .$$

Thus (1) follows. Equation (1) is called *Parseval's equation.* We have used that

$$\bar{\hat{f}}(k) = \int_{\mathbf{R}} \bar{f}(x) e^{-ikx} dx.$$

Chapter 15

Complex Analysis

Problem 1.   Let

$$z = x + iy, \quad \bar{z} = x - iy$$

where x, $y \in \mathbf{R}$. (i) Show that

$$\frac{\partial}{\partial z} = \frac{1}{2}\left(\frac{\partial}{\partial x} - i\frac{\partial}{\partial y}\right), \tag{1a}$$

$$\frac{\partial}{\partial \bar{z}} = \frac{1}{2}\left(\frac{\partial}{\partial x} + i\frac{\partial}{\partial y}\right). \tag{1b}$$

(ii) Find $dz \wedge d\bar{z}$, where $\wedge$ denotes the exterior product.

Solution.   (i) Let f be a smooth function of x and y ($x, y \in \mathbf{R}$). Then

$$df = \frac{\partial f}{\partial x}dx + \frac{\partial f}{\partial y}dy. \tag{2}$$

Since

$$dz = d(x + iy) = dx + idy, \quad d\bar{z} = d(x - iy) = dx - idy,$$

we obtain

$$dx = \frac{1}{2}(dz + d\bar{z}), \tag{3a}$$

$$dy = \frac{1}{2i}(dz - d\bar{z}). \tag{3b}$$

146

nserting (3a) and (3b) into (2) yields

$$df = \frac{\partial f}{\partial x}\frac{1}{2}(dz + d\bar{z}) + \frac{\partial f}{\partial y}\frac{1}{2i}(dz - d\bar{z})$$

$$= \frac{1}{2}\left(\frac{\partial f}{\partial x} - i\frac{\partial f}{\partial y}\right)dz + \frac{1}{2}\left(\frac{\partial f}{\partial x} + i\frac{\partial f}{\partial y}\right)d\bar{z}.$$

ince

$$df = \frac{\partial f}{\partial z}dz + \frac{\partial f}{\partial \bar{z}}d\bar{z},$$

e find (1a) and (1b).

ii) Since

$$dx \wedge dx = 0, \quad dy \wedge dy = 0, \quad dx \wedge dy = -dy \wedge dx,$$

e obtain

$$dz \wedge d\bar{z} = -2i dx \wedge dy.$$

Problem 2.   Discuss the mapping of the periodic strip

$$-\pi < \Re(z) \le +\pi$$

y the function

$$w(z) = \sin(z).$$

ind the images of the straight lines

$$\Re(z) = \text{const}$$

ad of the straight segments

$$\Im(z) = \text{const}.$$

olution.   Since

$$z = x + iy,$$

$$w(z(x,y)) = u(x,y) + iv(x,y),$$

here u and v are real-valued functions and

$$\sin z = \sin(x + iy) = \frac{e^y + e^{-y}}{2}\sin x + i\frac{e^y - e^{-y}}{2}\cos x,$$

we obtain

$$u(x, y) = \frac{e^y + e^{-y}}{2} \sin x \,,$$

$$v(x, y) = \frac{e^y - e^{-y}}{2} \cos x \,.$$

Let y_0 be a fixed positive number. If $y = y_0$ and x increases from $-\pi$ to π, then the function w describes an ellipse with foci at $\pm i$, since

$$\frac{u^2}{C^2} = \sin^2 x \,,$$

$$\frac{v^2}{D^2} = \cos^2 x \,,$$

and

$$\sin^2 x + \cos^2 x = 1$$

with

$$C = \frac{e^y + e^{-y}}{2} \,, \quad D = \frac{e^y - e^{-y}}{2} \,.$$

The ellipse is described once in the clockwise direction starting at its lowest point. For $y = -y_0$ we obtain the same ellipse described in the opposite direction starting at its highest point. Through each point of the plane (except those along the real segment $[-1, 1]$) passes exactly one such ellipse. If

$$y_0 = 0 \,,$$

then

$$u(x, y_0 = 0) = \sin x \,,$$

$$v(x, y_0 = 0) = 0 \,.$$

Consequently, we describe the segment $[-1, 1]$ twice (from 0 to -1, to 0 to 1, to 0). To obtain the complete image of the periodic strip, we take two copies of the w plane, cut one along the positive imaginary axis, the other along the negative imaginary axis and both along the real segment $[-1, 1]$. If we join the banks of the horizontal slits crosswise, we find the two-sheeted Riemann surface which is the one-to-one image of the periodic strips. The lines

$$\Re(z) = \text{const}$$

are taken into hyperbolas orthogonal to the ellipses.

Problem 3.   Calculate

$$\int_0^\infty \frac{x \sin(ax)}{x^2 + \lambda^2} dx$$

where a, $\lambda^2 \in \mathbf{R}$ and $\lambda^2 > 0$.

Solution.   We solve the problem by considering the integration in the complex plane. First we notice that

$$\int_0^\infty \frac{x \sin(ax)}{x^2 + \lambda^2} dx = \frac{1}{2} \int_{-\infty}^{+\infty} \frac{x \sin(ax)}{x^2 + \lambda^2} dx = \frac{1}{2} \Im \int_{-\infty}^{+\infty} \frac{x e^{iax}}{x^2 + \lambda^2} dx$$

where $\Im$ denotes the imaginary part. Now we consider

$$\int_C \frac{z e^{iaz}}{z^2 + \lambda^2} dz$$

where C is the contour consisting of the line along the x-axis from $-R$ to $+R$ and the semicircle Γ above the x-axis having this line as diameter. Thus we have

$$\int_C \frac{z e^{iaz}}{z^2 + \lambda^2} dz = \int_{-R}^{+R} \frac{x e^{iax}}{x^2 + \lambda^2} dx + \int_\Gamma \frac{R e^{i\phi} e^{iaR e^{i\phi}}}{R^2 e^{2i\phi} + \lambda^2} R e^{i\phi} i d\phi \qquad (1)$$

since

$$z = R e^{i\phi}$$

and

$$dz = R e^{i\phi} i d\phi.$$

Now

$$z^2 + \lambda^2 \equiv (z + i\lambda)(z - i\lambda)$$

where only the pole at $z = i\lambda$ lies inside C. The left-hand side is calculated with the help of the *residue theorem*. The residue a_1 of a complex function at $z = a$, where $z = a$ is a pole of order k, is calculated as follows

$$a_{-1} := \lim_{z \to a} \frac{1}{(k-1)!} \frac{d^{k-1}}{dz^{k-1}} ((z-a)^k f(z)).$$

If $k = 1$ (simple pole) we have

$$a_{-1} = \lim_{z \to a} (z - a)f(z).$$

The *residue* is given by

$$\text{Res}_{z=i\lambda} \frac{ze^{iaz}}{z^2 + \lambda^2} := \lim_{z \to i\lambda} \frac{z(z - i\lambda)e^{iaz}}{(z + i\lambda)(z - i\lambda)} = \frac{i\lambda e^{iai\lambda}}{2i\lambda} = \frac{1}{2}e^{-a\lambda}.$$

The residue theorem states that

$$\oint_C f(z)dz = 2\pi(a_{-1} + b_{-1} + c_{-1} + \cdots)$$

where $a_{-1}, b_{-1}, c_{-1}, \ldots$ are the residues inside C. Consequently,

$$\int_C \frac{ze^{iaz}}{z^2 + \lambda^2}dz = \pi i e^{-a\lambda}.$$

If $R \to \infty$, the second integral on the right hand side of (1) vanishes. Therefore

$$\int_{-\infty}^{+\infty} \frac{xe^{iax}}{x^2 + \lambda^2}dx = \pi i e^{-a\lambda}, \quad \Im \int_{-\infty}^{+\infty} \frac{xe^{iax}}{x^2 + \lambda^2}dx = \pi e^{-a\lambda}.$$

Finally

$$\int_0^{\infty} \frac{x \sin(ax)}{x^2 + \lambda^2}dx = \frac{\pi}{2}e^{-a\lambda}.$$

Problem 4.   (i) Calculate the Fourier transform $\hat{f}(\omega)$ of

$$f(t) = e^{-|t|}.$$

(ii) Calculate the inverse Fourier transform of $\hat{f}(\omega)$.

Solution.   (i) From the definition of the Fourier transform

$$\hat{f}(\omega) := \int_{-\infty}^{+\infty} f(t)e^{i\omega t}dt,$$

we obtain

$$\hat{f}(\omega) = \int_{-\infty}^{0} e^t e^{i\omega t}dt + \int_0^{\infty} e^{-t}e^{i\omega t}dt \equiv \int_{-\infty}^{0} e^{(i\omega+1)t}dt + \int_0^{\infty} e^{(i\omega-1)t}dt.$$

Integration yields

$$\hat{f}(\omega) = \frac{2}{1 + \omega^2}.$$

(ii) The inverse Fourier transform is given by

$$f(t) = \frac{1}{2\pi} \int_{-\infty}^{+\infty} \hat{f}(\omega) e^{-i\omega t} d\omega.$$

Therefore

$$f(t) = \frac{1}{\pi} \int_{-\infty}^{+\infty} \frac{1}{1+\omega^2} e^{-i\omega t} d\omega.$$

To calculate this integral we extend ω in the complex domain and consider

$$\frac{1}{\pi} \oint_C \frac{e^{-izt}}{1+z^2} dz.$$

Since the integrand has the two poles $z = \pm i$ and

$$\frac{1}{1+z^2} \equiv -\frac{1}{2i}\frac{1}{z+i} + \frac{1}{2i}\frac{1}{z-i},$$

we consider two paths C_1 and C_2. The path C_1 is the contour consisting of the line along the x-axis from $-R$ to R and the semicircle Γ above the x-axis having this line as diameter. The path C_2 is the contour consisting of the line along the x-axis from $-R$ to R and the semicircle Γ below the x-axis having this line as diameter. For the path C_1 we have

$$\frac{1}{\pi} \oint_{C_1} \frac{e^{-izt}}{1+z^2} dz \equiv -\frac{1}{2i\pi} \oint_{C_1} \frac{e^{-izt}}{z+i} dz + \frac{1}{2i\pi} \oint_{C_1} \frac{e^{-izt}}{z-i} dz.$$

The first integral on the right-hand side is equal to zero, since the pole $z = -i$ is not inside C_1. Consequently,

$$\frac{1}{\pi} \oint_{C_1} \frac{e^{-izt}}{1+z^2} dz = \frac{1}{2i\pi} \oint_{C_1} \frac{e^{-izt}}{z-i} dz.$$

Thus $f(t) = e^t$ for $t \in (-\infty, 0]$. Analogously, for the path C_2 we have

$$\frac{1}{\pi} \oint_{C_2} \frac{e^{-izt}}{1+z^2} dz = -\frac{1}{2i\pi} \oint_{C_2} \frac{e^{-izt}}{z+i} dz + \frac{1}{2i\pi} \oint_{C_2} \frac{e^{-izt}}{z-i} dz.$$

The second integral on the right-hand side is equal to zero, since the pole $z = i$ is not inside C_2. Consequently,

$$\frac{1}{\pi} \oint_{C_2} \frac{e^{-izt}}{1+z^2} dz = -\frac{1}{2i\pi} \oint_{C_2} \frac{e^{-izt}}{z+i} dz.$$

Thus $f(t) = e^{-t}$ for $t \in [0, \infty)$. Consequently,

$$f(t) = e^{-|t|}.$$

Problem 5.   Expand the functions

$$\sqrt{(z-1)(z-2)} \quad \text{for } |z| > 2, \tag{1a}$$

$$\ln\left(\frac{1}{1-z}\right) \quad \text{for } |z| > 1, \tag{1b}$$

into their *Laurent series*.

Solution.   For $|z| > 2$ we have

$$\sqrt{(z-1)(z-2)} = \pm z \left(1 - \frac{1}{z}\right)^{1/2} \left(1 - \frac{2}{z}\right)^{1/2}.$$

Applying the binomial expansions for the square root, we can write for $|z| > 2$

$$\pm z \left(1 - \binom{\frac{1}{2}}{1}\frac{1}{z} + \binom{\frac{1}{2}}{2}\frac{1}{z^2} - \cdots\right)\left(1 - \binom{\frac{1}{2}}{1}\frac{2}{z} + \binom{\frac{1}{2}}{2}\frac{2^2}{z^2} - \cdots\right)$$

$$= \pm\left(c_0 z - c_1 + \frac{c_2}{z} - \frac{c_3}{z^2} + \cdots\right)$$

where

$$c_n = \binom{\frac{1}{2}}{n} + 2\binom{\frac{1}{2}}{n-1}\binom{\frac{1}{2}}{1} + 2^2\binom{\frac{1}{2}}{n-2}\binom{\frac{1}{2}}{2} + \cdots + 2^n\binom{\frac{1}{2}}{n}.$$

The function (1b) does not have a Laurent expansion for $|z| > 1$, since the function is not single-valued there. If we put $z' = 1/z$, then we find

$$\ln\left(\frac{-z'}{1-z'}\right) = \ln(-z') + \ln\left(\frac{1}{1-z'}\right)$$

$$= \ln\left(-\frac{1}{z}\right) + \frac{1}{z} + \frac{1}{2z^2} + \frac{1}{3z^3} + \cdots.$$

Problem 6.   Assume that f is a meromorphic function in the finite plane C and has a finite number of poles $a_1, a_2, \ldots, a_m$ none of which is an integer

Assume, moreover, that

$$\lim_{z \to \infty} z f(z) = 0.$$

Then

$$\lim_{N \to \infty} \sum_{n=-N}^{n=N} f(n)$$

exists and

$$\lim_{N \to \infty} \sum_{n=-N}^{n=N} f(n) = -\sum_{k=1}^{m} \text{Res}(a_k; \pi f(z) \cot(\pi z))$$

where Res denotes the residue and $\cot(\pi z) \equiv \cos(\pi z)/\sin(\pi z)$. Calculate

$$\sum_{n=-\infty}^{\infty} \frac{1}{n^2 + n + 1}.$$

Solution. The function

$$f(z) = \frac{1}{z^2 + z + 1}$$

is meromorphic in the finite plane $\mathbf{C}$ and has two poles, because

$$z^2 + z + 1 \equiv (z - a_1)(z - a_2)$$

with

$$a_1 = -\frac{1 + i\sqrt{3}}{2}, \quad a_2 = \bar{a}_1.$$

The residue of

$$\frac{\pi \cot \pi z}{z^2 + z + 1}$$

at

$$z = a_1$$

is given by

$$\lim_{z \to a_1} \frac{(z - a_1)\pi \cot \pi z}{(z - a_1)(z - \bar{a}_1)} = \frac{\pi \cot(\pi a_1)}{a_1 - \bar{a}_1} = -\frac{i}{\sqrt{3}} \pi \cot\left(\frac{\pi(1 + i\sqrt{3})}{2}\right).$$

Similarly, the residue of

$$\frac{\pi \cot(\pi z)}{z^2 + z + 1}$$

at

$$z = \bar{a}_1$$

is given by

$$\lim_{z \to \bar{a}_1} \frac{(z - \bar{a}_1)\pi \cot(\pi z)}{(z - a_1)(z - \bar{a}_1)} = \frac{\pi \cot(\pi \bar{a}_1)}{\bar{a}_1 - a_1} = \frac{i}{\sqrt{3}}\pi \cot\left(\frac{\pi(1 - i\sqrt{3})}{2}\right).$$

Therefore

$$\sum_{n=-\infty}^{\infty} \frac{1}{n^2 + n + 1} = -\frac{2\pi i}{\sqrt{3}}\tan\left(\frac{i\pi\sqrt{3}}{2}\right) = \frac{2\pi}{\sqrt{3}}\tanh\left(\frac{\pi\sqrt{3}}{2}\right)$$

where we have used the identity

$$\cot\left(\frac{\pi}{2} + \alpha\right) \equiv -\tan\alpha.$$

Problem 7.   A function which is analytic everywhere in the finite complex plane (i.e. everywhere except at ∞) is called an *entire function*. For example, the functions $\sin z$, e^z and z^4 are entire functions. An entire function can be represented by a Taylor series which has an infinite radius of convergence. Conversely, if a power series has an infinite radius of convergence, it represents an entire function.

Solve the linear differential equation

$$\left(\frac{d}{dz} - \frac{\epsilon + \kappa^2}{z + \kappa} + \kappa\right) w(z) = 0 \tag{1}$$

where κ and ϵ are real constants. Impose the condition that w is an entire function.

Remark.   *The differential equation* (1) *is the* Bargmann *representation of the* displaced harmonic oscillator.

Solution.   Method 1: Using the power series ansatz

$$w(z) = \sum_{n=0}^{\infty} b_n(z + \kappa)^{n+s}$$

yields

$$s = \epsilon + \kappa^2$$

and the recurrence relation for the expansion coefficients b_n's is given by

$$b_n = \frac{(-\kappa)}{n} b_{n-1}.$$

This recurrence relation can easily be solved

$$b_n = \frac{(-\kappa)^n}{\Gamma(n+1)} b_0$$

where Γ denotes the gamma function, i.e.

$$\Gamma(n+1) = n!$$

with $n = 0, 1, 2, \ldots$. The series $\sum_n b_n z^n$ therefore converges for all z and

$$w(z) = b_0 (z+\kappa)^{\epsilon+\kappa^2} \sum_{n=0}^{\infty} \frac{(-\kappa)^n}{\Gamma(n+1)} (z+\kappa)^n = b_0 (z+\kappa)^{\epsilon+\kappa^2} e^{-\kappa(z+\kappa)}.$$

Imposing the condition that w is an entire function leads to

$$\epsilon + \kappa^2 = n$$

where $n = 0, 1, 2, \ldots$. In physics this is called a quantization condition.

Method 2: Equation (1) can be directly integrated since

$$\frac{dw}{dz} = \left(\frac{\epsilon+\kappa^2}{z+\kappa} - \kappa \right) w(z).$$

It follows that

$$\int \frac{dw}{w} = \int \left(\frac{\epsilon+\kappa^2}{z+\kappa} - \kappa \right) dz.$$

We find

$$w(z) = C(z+\kappa)^{\epsilon+\kappa^2} e^{-\kappa z}.$$

The requirement that w be entire implies that there be no branch-point singularities at

$$z = -\kappa$$

and therefore

$$\epsilon + \kappa^2 = n.$$

Problem 8.   Assume $Ox_1x_2x_3$ is the system of rectangular coordinates whose axes Ox_1, Ox_2 coincide with the real and imaginary axes Ox, Oy of the complex plane **C**. Assume, moreover, that the ray emanating from the north pole $N(0, 0, 1)$ of the unit sphere S^2

$$x_1^2 + x_2^2 + x_3^2 = 1 \tag{1}$$

and intersecting S^2 at $A(x_1, x_2, x_3)$ intersects **C** at the point z. The point $z = x + iy$ is then called the *stereographic projection* of $A(x_1, x_2, x_3)$ whereas A is called the spherical image of z. The stereographic projection is given by

$$x_1 = \frac{z + \bar{z}}{1 + |z|^2}, \quad x_2 = \frac{z - \bar{z}}{i(1 + |z|^2)}, \quad x_3 = \frac{|z|^2 - 1}{1 + |z|^2}$$

and

$$z = \frac{x_1 + ix_2}{1 - x_3}.$$

(i) Find the spherical images of $e^{i\alpha}$ where $\alpha \in \mathbf{R}$.

(ii) Let θ, ϕ be the geographical latitude and longitude of A respectively. Show that the stereographic projection of A has the representation

$$z = e^{i\phi} \tan\left(\frac{1}{4}\pi + \frac{1}{2}\theta\right). \tag{2}$$

(iii) The distance $\sigma(z_1, z_2)$ between two points on S^2 whose stereographic projections are z_1, z_2 is called the *spherical distance* or *chordal distance* between z_1 and z_2. Show that

$$\sigma(z_1, z_2) = \frac{2|z_1 - z_2|}{\sqrt{(1 + |z_1|^2)(1 + |z_2|^2)}} \tag{3}$$

is the Euclidean distance of the images of z_1 and z_2.

Solution.   (i) Since $z = e^{i\alpha}$ we have $\bar{z} = e^{-i\alpha}$ and $z\bar{z} = 1$. Therefore

$$z + \bar{z} = 2\cos\alpha, \quad z - \bar{z} = 2i\sin\alpha.$$

It follows that

$$x_1 = \cos\alpha, \quad x_2 = \sin\alpha, \quad x_3 = 0.$$

(ii) Since

$$x_1(\theta, \phi) = \sin\theta\cos\phi$$
$$x_2(\theta, \phi) = \sin\theta\sin\phi,$$
$$x_3(\theta, \phi) = \cos\theta,$$

we obtain

$$z = \frac{x_1 + ix_2}{1 - x_3} = \frac{\sin\theta\cos\phi + i\sin\theta\sin\phi}{1 - \cos\theta} = \frac{e^{i\phi}\sin\theta}{1 - \cos\theta}.$$

Obviously, (2) follows.

(iii) Let **x** and **y** be two points on the unit sphere (1). Then

$$x_1^2 + x_2^2 + x_3^2 = 1, \quad y_1^2 + y_2^2 + y_3^2 = 1. \tag{4}$$

It follows that

$$\|\mathbf{x} - \mathbf{y}\| = \sqrt{(x_1 - y_1)^2 + (x_2 - y_2)^2 + (x_3 - y_3)^2}$$
$$= \sqrt{2 - 2x_1y_1 - 2x_2y_2 - 2x_3y_3}$$

where we have used (4). Under the stereographic projection we have

$$z_1 \to (x_1, x_2, x_3), \quad z_2 \to (y_1, y_2, y_3).$$

Now

$$1 + |z_1|^2 = \frac{2}{1 - x_3}, \quad 1 + |z_2|^2 = \frac{2}{1 - y_3} \tag{5}$$

and

$$|z_1 - z_2| = \frac{\sqrt{(1 - y_3)^2(1 - x_3^2) + (1 - x_3)^2(1 - y_3^2) - 2(1 - x_3)(1 - y_3)(x_1y_1 + x_2y_2)}}{(1 - x_3)(1 - y_3)}. \tag{6}$$

Inserting (5) and (6) into (3) leads to

$$\sigma(z_1, z_2) = \|\mathbf{x} - \mathbf{y}\|.$$

Remark. *The stereographic projection can be extended to arbitrary dimensions. Let S^n denote the unit n-sphere in $\mathbf{R}^{n+1}$ and let $q = (0, \ldots, 0, 1)$ denote the "north pole" of S^n. Let*

$$f : \mathbf{R}^n \to S^n$$

be the map which sends each $p \in \mathbf{R}^n$ into the point different from q where
the line through $(p, 0) \in \mathbf{R}^{n+1}$ and q cuts S^n. Since

$$\alpha(t) = t(p, 0) + (1 - t)q = (tp, 1 - t)$$

is a parametrization of the line through $(p, 0)$ and q, and since

$$\|\alpha(t)\| = 1$$

if and only if $t = 0$ or

$$t = \frac{2}{\|p\|^2 + 1},$$

the map f is given by

$$f(x_1, x_2, \ldots, x_n) = \frac{(2x_1, 2x_2, \ldots, 2x_n, x_1^2 + x_2^2 + \cdots + x_n^2 - 1)}{x_1^2 + x_2^2 + \cdots + x_n^2 + 1}.$$

The mapping f is a parametrized surface which maps $\mathbf{R}^n$ one-to-one onto
$S^n \setminus \{q\}$. The chart f^{-1} is called the stereographic projection from $S^n \setminus \{q\}$
onto the equatorial hyperplane.

Problem 9.   A *Möbius transformation* is a function of the form

$$w(z) = \frac{az + b}{cz + d}$$

where a, b, c, d are complex numbers and $ad - bc \neq 0$. Find the Möbius
transformation that carries the point 1 to i, the point 0 to 1 and the point
i to ∞.

Solution.   Since 0 goes to 1, we must have $b/d = 1$. Thus

$$w(z) = \frac{az + b}{cz + b}.$$

Since i goes to ∞, $z = i$ must make the denominator zero. Hence we have
$ci + b = 0$, $b = -ic$, and therefore

$$w(z) = \frac{az - ic}{cz - ic}.$$

Finally, since 1 goes to i

$$i = \frac{a - ic}{c - ic}.$$

Thus $a = c(1 + 2i)$. The coefficient c is at our disposal, with $c \neq 0$. Let $c = 1$. Then we arrive at

$$w(z) = \frac{(1 + 2i)z - i}{z - i} \, .$$

Remark. *A Möbius transformation is a one-to-one map of the extended z plane to the extended w plane, with $w = 0$ when $az + b = 0$ and $w = \infty$ when $cz + d = 0$. The transformation is a composition of magnifications, rotations, translations, and a reciprocal transformation $w(z) = 1/z$.*

Problem 10. Give the domain in the complex plane where the infinite series

$$\sum_{n=1}^{\infty} \frac{1}{n^2} \exp\left(\frac{nz}{z - 2}\right) \tag{1}$$

converges.

Solution. Let

$$f(z) = \exp\left(\frac{z}{z - 2}\right) \, .$$

The series (1) can then be cast into the form

$$\sum_{n=1}^{\infty} \frac{1}{n^2} (f(z))^n \, .$$

Thus using the theory of power series, it converges if and only if $|f(z)| \leq 1$. This inequality holds when

$$\Re \frac{z}{z - 2} \leq 0 \, .$$

Thus we have to find the region which is sent into the closed left half-plane by the linear fractional map

$$h(z) = \frac{z}{z - 2} \, .$$

The inverse of this map is the map

$$g(z) = \frac{2z}{z-1}.$$

Since $g(0) = 0$ and $g(\infty) = 2$, the image of the imaginary axis under g is a circle passing through the points 0 and 2. We can conclude that $|f(z)| \leq 1$ if and only if $|z - 1| \leq 1$ and $z \neq 2$, which is the region of convergence of the series (1).

Chapter 16

Special Functions

Problem 1.   A generating function $F(x,t)$ of the *Hermite polynomial* $H_n(x)$ is given by

$$F(x,t) := e^{x^2-(t-x)^2} = \sum_{k=0}^{\infty} H_k(x)\frac{t^k}{k!}$$

where $n = 0, 1, 2, \ldots$ and $t, x \in \mathbf{R}$.

(i) Express $H_n(x)$ as a contour integral.

(ii) Prove that $H_n(x)$ satisfies *Hermite's differential equation*

$$\frac{d^2 H_n}{dx^2} - 2x\frac{dH_n}{dx} + 2nH_n = 0.$$

(iii) Show that

$$\frac{dH_n}{dx} = 2nH_{n-1}.$$

Solution.   (i) We know that

$$\oint \frac{1}{z^n}dz = \begin{cases} 2\pi i & \text{if } n = 1, \\ 0 & \text{otherwise} \end{cases} \tag{1}$$

where the integration is performed over a closed contour in the complex z-plane enclosing the origin and $n \in \mathbf{Z}$. Dividing the generating equation $F(x,t)$ by t^{n+1} and integrating over a closed contour in the complex t-plane

161

enclosing the origin, we obtain

$$H_n(x) = \frac{n!}{2\pi i} \oint \frac{\exp(x^2 - (t-x)^2)}{t^{n+1}} dt \tag{2}$$

where we have used (1).

(ii) One forms the quantity

$$\frac{\partial^2 F}{\partial x^2} + 2t\frac{\partial F}{\partial t} - 2x\frac{\partial F}{\partial x}$$

and verifies that it is identically zero. Using the expansion for F in terms of H_n, this identity takes the form

$$\sum_{n=0}^{\infty} \frac{(H_n''(x) - 2xH_n'(x) + 2nH_n(x))t^n}{n!} = 0$$

for all t, where $H_n' \equiv dH_n/dx$. It follows that

$$H_n''(x) - 2xH_n'(x) + 2nH_n(x) = 0.$$

(iii) Differentiating the integral representation for $H_n(x)$ given by (2) with respect to x we obtain

$$\frac{dH_n}{dx} = \frac{2n((n-1)!)}{2\pi i} \oint \frac{\exp(x^2 - (t-x)^2)}{t^n} dt.$$

Taking into account (2) we find that

$$\frac{dH_n}{dx} = 2nH_{n-1}.$$

Problem 2.   A generating function for the *Legendre polynomials* $P_l(x)$ is given by

$$G(x,r) := \frac{1}{(1 - 2xr + r^2)^{1/2}} = \sum_{l=0}^{\infty} r^l P_l(x)$$

where $x = \cos\theta$ and $|r| < 1$. Prove that $P_l(x)$ satisfies the linear differential relation

$$xP_l'(x) = P_{l-1}'(x) + lP_l(x) \tag{1}$$

where $P_l'(x) \equiv dP_l/dx$.

Solution. Differentiating G with respect to r yields

$$\frac{(x-r)}{(1-2xr+r^2)^{3/2}} = \sum_{l=0}^{\infty} l r^{l-1} P_l(x)\,.\qquad(2)$$

Differentiating G with respect to x gives

$$\frac{r}{(1-2xr+r^2)^{3/2}} = \sum_{l=0}^{\infty} r^l P_l'(x)\,.\qquad(3)$$

Eliminating $(1-2xr+r^2)^{-3/2}$ between (2) and (3) leads to

$$\sum_{l=0}^{\infty}(x-r)r^l P_l'(x) = \sum_{l=0}^{\infty} l r^l P_l(x)\,,$$

which, upon equating coefficients of equal powers of r, gives (1).

Remark. *The first three Legendre polynomials are given by*

$$P_0(x) = 1\,,\quad P_1(x) = x\,,\quad P_2(x) = \frac{1}{2}(3x^2 - 1)\,.$$

Problem 3. Let

$$T_{n+1}(x) - 2x T_n(x) + T_{n-1}(x) = 0\qquad(1)$$

where $n = 1, 2, \ldots$,

$$T_0(x) = 1\,,\quad T_1(x) = x\qquad(2)$$

and x is a fixed parameter. Solve the linear difference equation (1) where the initial values are given by (2).

Remark. *The quantities $T_n(x)$ are called Chebyshev polynomials of the first kind.*

Solution. Since (1) is a linear difference equation with constant coefficients it can be solved with the ansatz

$$T_n(x) := A(r(x))^n$$

where A is a constant. Inserting this ansatz into (1) yields

$$r^{n+1} - 2x r^n + r^{n-1} = 0$$

or

$$r^2 - 2x r + 1 = 0\,.\qquad(3)$$

The solution to (3) is given by

$$r_{1,2}(x) = x \pm \sqrt{x^2 - 1}.$$

Thus the solution of the difference equation (1) takes the form

$$T_n(x) = A(x + \sqrt{x^2 - 1})^n + B(x - \sqrt{x^2 - 1})^n$$

where A and B are the constants of integration and $n = 0, 1, 2, \ldots$. Imposing the initial conditions (2) yields

$$T_n(x) = \frac{1}{2}(x + \sqrt{x^2 - 1})^n + \frac{1}{2}(x - \sqrt{x^2 - 1})^n.$$

The solution can also be written as

$$T_n(x) = \cos(n \arccos x).$$

Problem 4.   The *metric tensor field* of the two-dimensional Euclidean space is given by

$$g = dx \otimes dx + dy \otimes dy. \tag{1}$$

Calculate the arc length of an *ellipse*.

Solution.   Let $a > b > 0$. Then the equation of the ellipse in parametric form is given by

$$x(\phi) = a \sin \phi, \quad y(\phi) = b \cos \phi$$

where $0 \le \phi < 2\pi$. Since

$$dx = a \cos(\phi)d\phi, \quad dy = -b \sin(\phi)d\phi,$$

we obtain

$$dx \otimes dx = a^2 \cos^2(\phi)d\phi \otimes d\phi, \quad dy \otimes dy = b^2 \sin^2(\phi)d\phi \otimes d\phi.$$

Consequently, we obtain from (1)

$$g = (a^2 \cos^2 \phi + b^2 \sin^2 \phi)d\phi \otimes d\phi.$$

Therefore the *line element* is given by

$$\left(\frac{ds}{d\phi}\right)^2 = a^2 \cos^2 \phi + b^2 \sin^2 \phi.$$

Using the identity

$$\sin^2 \phi + \cos^2 \phi \equiv 1,$$

we arrive at

$$\int_0^{2\pi} ds = \int_0^{2\pi} \sqrt{a^2 - (a^2 - b^2)\sin^2 \phi} \, d\phi$$

or

$$\int_0^{2\pi} ds = a \int_0^{2\pi} \sqrt{1 - \frac{a^2 - b^2}{a^2} \sin^2 \phi} \, d\phi.$$

This is a *complete elliptic integral of the second kind,* where

$$k^2 = \frac{a^2 - b^2}{a^2} =: e^2.$$

Here e is the *eccentricity* of the ellipse. We find

$$\int_0^{2\pi} ds = 4aE(k, \pi/2).$$

Remark. *The complete elliptic integral of the second kind can be given as*

$$E(k, \pi/2) = \int_0^{\pi/2} \sqrt{1 - k^2 \sin^2 \phi} \, d\phi.$$

Thus

$$E(k, \pi/2) = \frac{\pi}{2} \left(1 - \left(\frac{1}{2}\right)^2 k^2 - \left(\frac{1\cdot 3}{2\cdot 4}\right)^2 \frac{k^4}{3} - \left(\frac{1\cdot 3\cdot 5}{2\cdot 4\cdot 6}\right)^2 \frac{k^6}{5} - \cdots \right)$$

where $|k^2| < 1$. *We have used the Taylor expansion around* $x = 0$ *of the function*

$$f(x) = \sqrt{1 - x^2}$$

with $x^2 < 1$, *i.e.*

$$\sqrt{1 - x^2} = 1 - \frac{1}{2}x^2 - \frac{1}{2\cdot 4}x^4 - \frac{1\cdot 3}{2\cdot 4\cdot 6}x^6 - \cdots, \quad -1 < x \le 1.$$

Problem 5. (i) Let

$$\text{sn}^{-1}(s, k) := \int_0^{\arcsin s} \frac{dx}{\sqrt{1 - k^2 \sin^2 x}}$$

where $k \in [0, 1]$. Calculate

$$\frac{dsn^{-1}(s, k)}{ds} .$$

(ii) Let

$$cn^{-1}(s, k) := \int_0^{arccos\, s} \frac{dx}{\sqrt{1 - k^2 \sin^2 x}} .$$

Calculate

$$\frac{dcn^{-1}(s, k)}{ds} .$$

Solution.   We use *Leibniz's rule* for differentiation of integrals

$$\frac{d}{ds} \int_{f(s)}^{g(s)} F(x, s) dx = \int_{f(s)}^{g(s)} \frac{\partial F}{\partial s} dx + F(g(s), s) \frac{dg}{ds} - F(f(s), s) \frac{df}{ds} . \quad (1)$$

If F is independent of s, then (1) simplifies to

$$\frac{d}{ds} \int_{f(s)}^{g(s)} F(x) dx = F(g(s)) \frac{dg}{ds} - F(f(s)) \frac{df}{ds} .$$

(i) Applying this rule we obtain

$$\frac{dsn^{-1}(s, k)}{ds} = \frac{1}{\sqrt{1 - k^2 (\sin(\arcsin s))^2}} \frac{d}{ds} \arcsin s .$$

Since

$$\sin(\arcsin s) = s$$

and

$$\frac{d}{ds} \arcsin s = \frac{1}{\sqrt{1 - s^2}}$$

we obtain

$$\frac{dsn^{-1}(s, k)}{ds} = \frac{1}{\sqrt{1 - k^2 s^2}} \frac{1}{\sqrt{1 - s^2}} .$$

(ii) We find

$$\frac{dcn^{-1}(s, k)}{ds} = \frac{1}{\sqrt{1 - k^2 (\sin(\arccos s))^2}} \frac{d}{ds} \arccos s .$$

Since

$$\sin(\arccos s) = \sqrt{1 - s^2}$$

and

$$\frac{d}{ds} \arccos s = -\frac{1}{\sqrt{1 - s^2}},$$

we obtain

$$\frac{d\text{cn}^{-1}(s, k)}{ds} = -\frac{1}{\sqrt{1 - k^2(1 - s^2)}} \frac{1}{\sqrt{1 - s^2}}.$$

Problem 6. Consider the function

$$\sigma(z; \omega_1, \omega_2) = z \prod_k' \left(1 - \frac{z}{\Omega_k}\right) \exp\left(\frac{z}{\Omega_k} + \frac{1}{2}\left(\frac{z}{\Omega_k}\right)^2\right)$$

with

$$\Omega_k = m_k \omega_1 + n_k \omega_2$$

where m_k, n_k is the sequence of all pairs of integers. The prime after the product sign indicates that the pair $(0, 0)$ should be omitted. ω_1 and ω_2 are two complex numbers with

$$\Im(\omega_1/\omega_2) \neq 0.$$

In the following the dependence on ω_1 and ω_2 will be omitted. The logarithmic derivative $\sigma'(z)/\sigma(z)$ where $\sigma'(z) \equiv d\sigma/dz$ is the meromorphic function $\zeta(z)$ of Weierstrass. The function

$$\wp(z) = -\zeta'(z)$$

is a meromorphic, doubly periodic (or elliptic) function with periods ω_1, ω_2 whose only singularities are double poles $m\omega_1 + n\omega_2$. We find

$$\wp(z; \omega_1, \omega_2) = \frac{1}{z^2} + \sum_k' \left(\frac{1}{(z - \Omega_k)^2} - \frac{1}{(\Omega_k)^2}\right).$$

Remark. *The function $\wp$ is called Weierstrass function $\wp$.*

(i) Show that

$$\sigma(z) = z + c_5 z^5 + c_7 z^7 + \cdots. \tag{1}$$

(ii) Show that

$$\frac{\sigma(2u)}{\sigma^4(u)} = -\wp'(u),\tag{2a}$$

$$2\zeta(2u) - 4\zeta(u) = \frac{\wp''(u)}{\wp'(u)}.\tag{2b}$$

Solution.   (i) Grouping the factors corresponding to the pairs (m,n), $(-m,-n)$ we find

$$\sigma(z) = z\prod_k{}' \left(1 - \frac{z}{\Omega_k^2}\right)\exp(z^2/\Omega_k^2).$$

Since every factor has the form

$$1 - z^4/\Omega_k^2$$

we find (1).

(ii) The function

$$\frac{\sigma(2u)}{\sigma^4(u)}$$

is doubly periodic and has poles of order three at $m\omega_1 + n\omega_2$. Furthermore the function is equal to

$$2\frac{1}{u^3} + Au + \cdots$$

near the origin. Consequently,

$$\wp'(u) + \frac{\sigma(2u)}{\sigma^4(u)} = 0\tag{3}$$

since the left hand side has only removable singularities.

(iii) Calculating the logarithmic derivatives

$$(\ln f)' = \frac{f'}{f}$$

of both sides of (3) we find (2b).

Problem 7. Let $a(q)$ be a meromorphic function. Let r, k, $l = 1, 2, \ldots, N$. Assume that a satisfies the following equation

$$\frac{a(q_k - q_r)a'(q_r - q_l) - a'(q_k - q_r)a(q_r - q_l)}{a(q_k - q_l)} = g(q_k - q_r) - g(q_r - q_l) \quad (1)$$

where a' denotes differentiation with respect to the arguments and $k \neq l$. Let

$$U(q) = a^2(q).$$

Show that

$$[U(x)U'(y) - U'(x)U(y)] + [U(y)U'(z) - U'(y)U(z)]$$
$$+ [U(z)U'(x) - U'(z)U(x)] = 0$$

where

$$x + y + z = (q_k - q_r) + (q_r - q_l) + (q_l - q_k) = 0.$$

Solution. The permutations of (k, r, l) are given by

$$(k, r, l), (k, l, r), (r, k, l), (r, l, k), (l, k, r), (l, r, k).$$

Adding all Eq. (1) corresponding to all permutations of (k, r, l) we find

$$\frac{a(q_r - q_l)a'(q_l - q_k) - a'(q_r - q_l)a(q_l - q_k)}{a(q_k - q_r)}$$

$$+ \frac{a(q_l - q_k)a'(q_k - q_r) - a'(q_l - q_k)a(q_k - q_r)}{a(q_r - q_l)}$$

$$+ \frac{a(q_k - q_r)a'(q_r - q_l) - a'(q_k - q_r)a(q_r - q_l)}{a(q_l - q_k)} = 0.$$

Therefore the equation for U follows.

Remark. *This equation is the addition formula for the Weierstrass function.*

Problem 8. The *formula of Rodrigues* for the *Legendre polynomials* is given by

$$P_n(x) := \frac{1}{2^n n!} \frac{d^n}{dx^n} (x^2 - 1)^n.$$

Let f be an analytic function.

(i) Show that the integral

$$I = \int_{-1}^{1} f(x)P_n(x)dx$$

can be brought into the form

$$I = \frac{(-1)^n}{2^n n!} \int_{-1}^{1} (x^2 - 1)^n \frac{d^n}{dx^n} f(x)dx. \tag{1}$$

(ii) Show that

$$\int_{-1}^{1} P_m(x)P_n(x)dx = 0, \quad m \neq n. \tag{2}$$

Solution.   (i) Applying integration by parts to (1), we obtain

$$I = \frac{1}{2^n n!} \left| f(x)\frac{d^{n-1}}{dx^{n-1}}(x^2 - 1)^n \right|_{-1}^{+1} - \frac{1}{2^n n!} \int_{-1}^{1} \frac{df}{dx} \left(\frac{d^{n-1}}{dx^{n-1}}(x^2 - 1)^n dx \right). \tag{3}$$

Since $x^2 - 1$ is equal to zero at 1 and -1 we find that the first term on the right-hand side of (3) vanishes. Repeating this process of integration by parts we obtain (1).

(ii) Let $f(x) = P_m(x)$ with $m < n$. Then $d^n f/dx^n = 0$ and therefore (2) follows.

Chapter 17

Inequalities

Problem 1.   (i) Let a, $b \in \mathbf{R}$. Show that

$$a^2 + b^2 \geq 2ab.$$

(ii) Let a, b be two non-negative numbers. Show that

$$\frac{1}{2}(a + b) \geq \sqrt{ab}.$$

Solution.   (i) Since

$$(a - b)^2 \geq 0,$$

we find

$$a^2 + b^2 - 2ab \geq 0.$$

It follows that

$$a^2 + b^2 \geq 2ab.$$

(ii) Since

$$(\sqrt{a} - \sqrt{b})^2 \geq 0,$$

we have

$$a + b - 2\sqrt{ab} \geq 0.$$

Therefore

$$a + b \geq 2\sqrt{ab}.$$

Problem 2.   Let $a_1, a_2, \ldots, a_n, b_1, b_2, \ldots, b_n \in \mathbf{R}$.

(i) Show that

$$(a_1 b_1 + a_2 b_2 + \cdots + a_n b_n)^2 \leq (a_1^2 + a_2^2 + \cdots + a_n^2)(b_1^2 + b_2^2 + \cdots + b_n^2). \qquad (1)$$

(ii) Show that

$$\sqrt{(a_1 + b_1)^2 + \cdots + (a_n + b_n)^2} \leq \sqrt{a_1^2 + \cdots + a_n^2} + \sqrt{b_1^2 + \cdots + b_n^2}. \qquad (2)$$

Solution.   (i) If $a_1 = a_2 = \cdots = a_n = 0$, then inequality (1) is satisfied. Let us now assume that at least one of the a_j's is nonzero. Let ϵ be an arbitrary real number. Then we obviously have

$$(a_1 \epsilon + b_1)^2 + (a_2 \epsilon + b_2)^2 + \cdots + (a_n \epsilon + b_n)^2 \geq 0. \qquad (3)$$

Let

$$A^2 := a_1^2 + a_2^2 + \cdots + a_n^2,$$
$$B^2 := b_1^2 + b_2^2 + \cdots + b_n^2,$$
$$C := a_1 b_1 + a_2 b_2 + \cdots + a_n b_n.$$

Then from (3) we have

$$A^2 \epsilon^2 + 2C\epsilon + B^2 \geq 0. \qquad (4)$$

The left-hand side is a smooth function of ϵ, say

$$f(\epsilon) := A^2 \epsilon^2 + 2C\epsilon + B^2.$$

Since

$$\frac{df}{d\epsilon} = 2A^2 \epsilon + 2C, \qquad \frac{d^2 f}{d\epsilon^2} = 2A^2 > 0,$$

we find that f has a minimum at

$$\epsilon = -\frac{C}{A^2}.$$

Inserting ϵ into the left-hand side of inequality (4) we find

$$\frac{A^2 C^2}{A^4} - \frac{2C^2}{A^2} + B^2 \geq 0$$

or

$$A^2 B^2 \geq C^2 .$$

This is inequality (1).

(ii) Obviously

$$0 \leq \sum_{j=1}^{n}(a_j + b_j)^2 \equiv \sum_{j=1}^{n} a_j^2 + \sum_{j=1}^{n} b_j^2 + 2\sum_{j=1}^{n} a_j b_j . \tag{5}$$

From (1) we have

$$a_1 b_1 + a_2 b_2 + \cdots + a_n b_n \equiv \sum_{j=1}^{n} a_j b_j \leq \sqrt{\sum_{j=1}^{n} a_j^2} \sqrt{\sum_{j=1}^{n} b_j^2} .$$

Therefore from (5) we obtain

$$\sum_{j=1}^{n}(a_j + b_j)^2 \leq \sum_{j=1}^{n} a_j^2 + \sum_{j=1}^{n} b_j^2 + 2\sqrt{\sum_{j=1}^{n} a_j^2} \sqrt{\sum_{j=1}^{n} b_j^2} .$$

Taking the square root on both sides leads to (2).

Problem 3. Let $x \geq 0$ and

$$f(x) = x \ln x - x + 1 . \tag{1}$$

Show that

$$f(x) \geq 0 . \tag{2}$$

Remark. *From L'Hospital's rule we find that $f(0) = 1$.*

Solution. From (1) we obtain

$$\frac{df}{dx} = \ln x \tag{3}$$

and

$$\frac{d^2 f}{dx^2} = \frac{1}{x} . \tag{4}$$

From (3) and (4) we find that the function f has one minimum at

$$x = 1$$

for $x \geq 0$, since $\ln 1 = 0$ and

$$\frac{d^2 f}{dx^2}\Big|_{x=1} = 1 > 0.$$

It follows that

$$f(1) = 0.$$

Therefore inequality (2) follows.

Remark.   *The inequality (2) is called the* Gibbs' inequality *and plays an important rôle in statistical physics.*

Problem 4.   Suppose that $f : \mathbf{R} \to \mathbf{R}$ is twice differentiable with

$$f''(x) \geq 0$$

for all x. Prove that for all a and b, $a < b$,

$$f\left(\frac{a+b}{2}\right) \leq \frac{f(a) + f(b)}{2}. \tag{1}$$

Solution.   By the mean-value theorem there is a number x_1 in $(a, \frac{1}{2}(a+b))$ such that

$$\frac{f(\frac{1}{2}(a+b)) - f(a)}{\frac{1}{2}(a+b) - a} = f'(x_1)$$

and a number x_2 in $(\frac{1}{2}(a+b), b)$ such that

$$\frac{f(b) - f(\frac{1}{2}(a+b))}{b - \frac{1}{2}(a+b)} = f'(x_2).$$

However

$$f''(x) \geq 0$$

for all x in (x_1, x_2), so f' is a nondecreasing function. Thus

$$f'(x_2) \geq f'(x_1)$$

or equivalently,

$$\frac{f(b) - f(\frac{1}{2}(a+b))}{b - a} \geq \frac{f(\frac{1}{2}(a+b)) - f(a)}{b - a}.$$

Thus (1) follows.

Problem 5. Let $a_j > 0$ for $j = 1, 2, \ldots, n$. The *arithmetic mean* of a_1, $a_2, \ldots, a_n$ is the number

$$\frac{a_1 + a_2 + \cdots + a_n}{n}$$

and the *geometric mean* of $a_1, a_2, \ldots, a_n$ is the number

$$(a_1 a_2 \cdots a_n)^{1/n}.$$

i) Show that

$$(a_1 a_2 \cdots a_n)^{1/n} \le \frac{a_1 + a_2 + \cdots + a_n}{n}. \tag{1}$$

ii) Consider a rectangular parallelepiped. Let the lengths of the three adjacent sides be a, b and c. Let A and V be the surface area and the volume, respectively. Prove that

$$A \ge 6V^{2/3}. \tag{2}$$

Solution. (i) Let $n = 2$. Then we have to prove that

$$(a_1 a_2)^{1/2} \le \frac{a_1 + a_2}{2}.$$

This inequality was proved in Problem 1. The proof for larger values of n can be handled by mathematical induction.

ii) Obviously

$$A = 2(ab + bc + ca), \quad V = abc.$$

Thus $V^2 = a^2 b^2 c^2 = (ab)(bc)(ca)$. Using the arithmetic mean geometric mean inequality (1), we obtain

$$V^2 \le \left(\frac{ab + bc + ca}{3} \right)^3 = \left(\frac{2(ab + bc + ca)}{6} \right)^3 = \left(\frac{A}{6} \right)^3.$$

Thus inequality (2) follows. If $ab = bc = ca$ or equivalently $a = b = c$ we obtain

$$6V^{2/3} = A.$$

Problem 6. Let $x \in \mathbf{R}$ and $x > 0$. Show that

$$\sqrt[e]{e} \ge \sqrt[x]{x}.$$

Solution.   Since

$$f(x) = \sqrt[x]{x} \equiv \exp\left(\frac{1}{x}\ln x\right), \qquad x > 0,$$

we find

$$\frac{df}{dx} = \frac{1 - \ln x}{x^2}\exp\left(\frac{1}{x}\ln x\right).$$

The condition for an extremum is

$$\frac{df}{dx} = 0.$$

It follows that

$$1 - \ln(x) = 0 \Rightarrow x = e \quad \text{(critical point)}.$$

Therefore

$$f(e) = e^{1/e} \approx 1.44466786101\ldots.$$

The second derivative is

$$\frac{df^2}{dx^2} = \left(\left(\frac{1 - \ln x}{x^2}\right)^2 + \frac{-x - (1 - \ln x)2x}{x^4}\right)\exp\left(\frac{1}{x}\ln x\right).$$

Setting $x = e$ and using $\ln e = 1$ yields

$$\left.\frac{d^2 f}{dx^2}\right|_{x=e} = \left(-\frac{1}{e^3}\right)\exp\left(\frac{1}{e}\right) < 0.$$

Thus f has a maximum at

$$x = e.$$

For $x = 0$ we find that f takes the form

$$f(x = 0) = 0 \quad \text{(L'Hospital's rule)}.$$

Furthermore we have

$$\lim_{x\to\infty}\sqrt[x]{x} = 1.$$

Chapter 18

Functional Analysis

Problem 1. Let A and B be two arbitrary $n \times n$ matrices over $\mathbf{R}$. Define

$$(A, B) := \operatorname{tr}(AB^T) \tag{1}$$

where B^T denotes the transpose of B.

i) Show that

$$(A, A) \geq 0, \tag{2a}$$

$$(A, B) = (B, A), \tag{2b}$$

$$(cA, B) = c(A, B), \tag{2c}$$

$$(A_1 + A_2, B) = (A_1, B) + (A_2, B), \tag{2d}$$

where $c \in \mathbf{R}$.

ii) Can the result be extended to infinite-dimensional matrices?

Solution. (i) We find

$$(A, A) = \operatorname{tr}(AA^T) = \sum_{j=1}^{n} \sum_{k=1}^{n} a_{jk} a_{jk}.$$

Since

$$a_{jk} a_{jk} \geq 0,$$

we obtain inequality (2a). Owing to

$$(A, B) = \mathrm{tr}(AB^T) = \sum_{j=1}^{n} \sum_{k=1}^{n} a_{jk} b_{jk}$$

and

$$(B, A) = \mathrm{tr}(BA^T) = \sum_{j=1}^{n} \sum_{k=1}^{n} b_{jk} a_{jk},$$

we obtain (2b). Since

$$(cA, B) = \mathrm{tr}(cAB^T) = c\,\mathrm{tr}(AB^T) = c(A, B),$$

we find (2c). To prove (2d), we recall that

$$\mathrm{tr}(X + Y) = \mathrm{tr}(X) + \mathrm{tr}(Y)$$

for $n \times n$ matrices X and Y. Therefore

$$(A_1 + A_2, B) = \mathrm{tr}((A_1 + A_2)B^T) = \mathrm{tr}(A_1 B^T) + \mathrm{tr}(A_2 B^T)$$
$$= (A_1, B) + (A_2, B).$$

Remark.   *Consequently, definition (1) defines a scalar product for the vector space of $n \times n$ matrices over the field of real numbers.*

(ii) The results can be extended when we impose the condition

$$\sum_{j=1}^{\infty} \sum_{k=1}^{\infty} |a_{jk}|^2 < \infty. \tag{3}$$

Remark.   *If an infinite-dimensional matrix A satisfies condition (3) we call A a* Hilbert–Schmidt operator. *The norm of a Hilbert–Schmidt operator A is given by*

$$\|A\| := \sqrt{\sum_{j=1}^{\infty} \sum_{k=1}^{\infty} |a_{jk}|^2}.$$

Problem 2.   Show that if $a < b$, and $n \in \mathbf{Z}$ the functions

$$f_n(x) = e^{inx}$$

are a linearly independent set on $[a, b]$.

Solution.   If $b - a \geq 2\pi$ and

$$\sum_{n=p}^{q} c_n e^{inx} = 0$$

on $[a, b]$, then the orthogonality relationships among the functions f_n, i.e.

$$\int_0^{2\pi} e^{i(n-m)x} dx = 2\pi \delta_{mn}$$

imply that all c_n are equal to zero. Here δ_{mn} denotes the Kronecker delta. If

$$0 < b - a < 2\pi$$

and

$$g(x) = \sum_{n=p}^{q} c_n e^{inx} = 0$$

on $[a, b]$, then g has an extension to an entire function on $\mathbf{C}$. Thus

$$g(x) = 0,$$

$$\sum_{n=p}^{q} c_n e^{inx} = 0$$

in $[0, 2\pi]$ and the previous argument shows all c_n are equal to zero.

Problem 3.   Let $L_2[0, 1]$ be the Hilbert space of the square-integrable functions in the Lebesgue sense over the unit interval $[0, 1]$. Let $f : [0, 1] \mapsto [0, 1]$ be

$$f(x) = \begin{cases} 2x & \text{for } 0 \leq x \leq \dfrac{1}{2}, \\ 2(1 - x) & \text{for } \dfrac{1}{2} < x \leq 1. \end{cases}$$

basis in $L_2[0, 1]$ is given by

$$\{u_n(x) := e^{2\pi i x n} : n \in \mathbf{Z}\}.$$

Obviously, $f \in L_2[0, 1]$. Then f can be expanded in a Fourier series with respect to u_n. Find the Fourier coefficients.

Solution.   The *Fourier expansion* of f is given by

$$f(x) = \sum_{n \in \mathbf{Z}} (f, u_n) u_n \equiv \sum_{n \in \mathbf{Z}} c_n u_n$$

where $(,)$ denotes the scalar product in the Hilbert space $L_2[0,1]$. The scalar product is defined by

$$(f, g) := \int_0^1 f(x) \bar{g}(x) dx$$

where $\bar{g}$ is the complex conjugate of g. Hence the expansion coefficients c_n are given by

$$c_n \equiv (f, u_n) := \int_0^1 f(x) \bar{u}_n(x) dx.$$

It follows that

$$c_n = \int_0^1 f(x) e^{-2\pi i x n} dx$$

or

$$c_n = 2 \int_0^{1/2} x e^{-2\pi i x n} dx + 2 \int_{1/2}^1 e^{-2\pi i x n} dx - 2 \int_{1/2}^1 x e^{-2\pi i x n} dx.$$

For $n = 0$ we find

$$c_0 = \frac{1}{2}.$$

For $a \neq 0$ we have

$$\int e^{ax} dx = \frac{1}{a} e^{ax} \tag{1a}$$

$$\int x e^{ax} dx = \frac{1}{a} e^{ax} x - \frac{1}{a^2} e^{ax}. \tag{1b}$$

Using (1a) and (1b) we find for $n \neq 0$ that

$$c_n = \frac{4}{a^2} (1 - (-1)^n)$$

where

$$a = -2\pi i n.$$

Problem 4. Let $L_2(\mathbf{R}^2)$ be the Hilbert space of the square-integrable functions over $\mathbf{R}^2$. Let $C^2(\mathbf{R}^2, \mathbf{R})$ be the set of $\mathbf{R}$ valued functions having two continuous derivatives on $\mathbf{R}^2$. Assume that

$$f, g \in C^2(\mathbf{R}^2, \mathbf{R}) \cap L_2(\mathbf{R}^2) \quad \Delta f, \Delta g \in L_2(\mathbf{R}^2)$$

and that

$$\nabla f, \nabla g \in L_2(\mathbf{R}^2).$$

This means each component of each vector is in $L_2(\mathbf{R}^2)$. Show that

$$\int_{\mathbf{R}^2} f(x, y) \Delta g(x, y) dx dy = \int_{\mathbf{R}^2} g(x, y) \Delta f(x, y) dx dy. \tag{1}$$

Solution. Owing to the assumptions we know that all functions are in the Hilbert space $L_2(\mathbf{R}^2)$. It follows that all integrals exist. We apply *Green's theorem* to solve the problem. Let P and Q be two real valued functions that are continuous on a region $E \subset \mathbf{R}^2$ bounded by a rectifiable Jordan curve Γ and such that $\partial P/\partial y$ and $\partial Q/\partial x$ exist and are bounded in their interior of E and that

$$\int \int_E \frac{\partial Q}{\partial x} dx dy \quad \text{and} \quad \int \int_E \frac{\partial P}{\partial y} dx dy$$

exist. Then, if Γ is positively oriented, the line integral exists and we have

$$\int_\Gamma (P dx + Q dy) = \int \int_E \left(\frac{\partial Q}{\partial x} - \frac{\partial P}{\partial y} \right) dx dy.$$

Green's theorem implies that for all positive R,

$$\int_{B(0,R)} (f \Delta g - g \Delta f) dx dy = \int_{\partial B(0,R)} \left(f \frac{\partial g}{\partial x} - g \frac{\partial f}{\partial x} \right) dy$$

$$+ \int_{\partial B(0,R)} \left(g \frac{\partial f}{\partial y} - f \frac{\partial g}{\partial y} \right) dx.$$

An estimate for the integrals in these equations is

$$\left| \int_{\partial B(0,R)} f \frac{\partial g}{\partial x} dy \right|$$

$$\leq \int_0^{2\pi} |f(R \cos \phi, R \sin \phi)| \left| \frac{\partial g(R \cos \phi, R \sin \phi)}{\partial x} \right| |R \sin \phi| d\phi = A(R)$$

where we have introduced polar coordinates. Let $\| \cdot \|_2$ be the norm of the Hilbert space $L_2(\mathbf{R}^2)$. Since

$$\int_{B(0,R)} \left| f(r\cos\phi, r\sin\phi) \frac{\partial g(r\cos\phi, r\sin\phi)}{\partial x} \right| r\,dr\,d\phi$$

$$= \int_0^R A(r) r\,dr \leq \|f\|_2 \|\partial g/\partial x\|_2 \,,$$

it follows that there is a sequence $\{R_n\}_{n=1}^\infty$ such that

$$R_n \to \infty$$

and

$$A(R_n) \to 0$$

as $n \to \infty$. In a similar manner the other three integrals may be estimated. Thus

$$\left| \int_{B(0,R_n)} (f\Delta g - g\Delta f)\,dx\,dy \right|$$

is dominated by the sum of four quantities, each of which tends to zero as $n \to \infty$. Consequently, (1) follows.

Problem 5.   (i) Let $I = [0,1]$ with Lebesgue measure and $x \in I$. Let the nth *Rademacher function* r_n be defined as

$$r_n(x) := \operatorname{sgn}(\sin(2^n \pi x))$$

where $n = 0, 1, 2, \ldots$ and sgn denotes the signum function. Show that the Rademacher functions r_n constitute a linearly independent set.

(ii) For $m \in \mathbf{N}$ there is in $\mathbf{N} \cup \{0\}$ a unique finite set

$$\{n_1, n_2, \ldots, n_{K_m}\}$$

such that

$$m = \sum_{k=1}^{K_m} 2^{n_k} \,.$$

The mth *Walsh function* is defined by

$$W_m(x) := \prod_{k=1}^{K_m} r_{n_k}(x)$$

where $x \in I \equiv [0,1]$. Show that $\{W_m\}_{m=1}^{\infty}$ is a complete orthogonal set in the Hilbert space $L_2(I)$. This is the Hilbert space of the square-integrable functions.

Solution. (i) The function r_n partitions $[0, 1]$ into three sets, $S_n^{\pm}$ and S_n^0, where

$$S_n^{\pm} = r_n^{-1}(\pm 1), \quad S_n^0 = r_n^{-1}(0).$$

Furthermore,

$$S_0^+ = (0,1), \quad S_0^- = \emptyset, \quad S_0^0 = \{0,1\}.$$

If $n \geq 1$ each of $S_n^{\pm}$ consists of 2^{n-1} disjoint intervals, each of length 2^{-n} and S_n^0 consists of their $2^n + 1$ endpoints. The intervals of S_n^+ alternate with those of S_n^-. If $m > n$ the intervals of $S_m^{\pm}$ equipartition those of $S_n^{\pm}$ from which the linear independence of the Rademacher functions r_n, $n = 1, 2, \ldots$ follows.

Remark. *Obviously, $\{r_n\}_n$ is an orthonormal set in the Hilbert space $L_2(I)$. However, the Rademacher functions do not form a basis in the Hilbert space $L_2(I)$.*

(ii) The result of (i) shows that

$$\{W_m\}_{m=1}^{\infty}$$

is an orthonormal set in the Hilbert space $L_2(I)$. Let

$$f \in (\{W_m\}_{m=1}^{\infty})^{\perp}$$

where $M^{\perp}$ is the orthogonal complement of M. The set $M^{\perp}$ denotes all vectors which are orthogonal to every vector in M. Then for all $N \in \mathbf{N}$ and all x,

$$F(x) = \int_0^1 f(t) \prod_{m=0}^{N} (1 + r_m(x) r_m(t)) dt = 0.$$

Induction shows that if $N > 1$ and

$$\frac{k}{2^m} \leq x \leq \frac{(k+1)}{2^m},$$

then

$$\prod_{m=0}^{N}(1+r_m(x)r_m(t)) \neq 0$$

if and only if

$$\frac{k}{2^N} \leq t \leq \frac{(k+1)}{2^N}.$$

Obviously

$$1 + r_2(x)r_2(t) \neq 0$$

if and only if x and t are in the same half of I,

$$(1 + r_2(x)r_2(t))(1 + r_3(x)r_3(t)) \neq 0$$

if and only if x and t are in the same quarter of I, etc. Hence

$$\int_{k/2^N}^{(k+1)/2^N} f(t)dt = 0$$

for all k, $N \in \mathbf{N}$. Consequently, $f(t) = 0$ almost everywhere. It follows that

$$\{W_m\}_{m=1}^{\infty}$$

is a complete orthornormal system. In other words the Walsh functions form a basis in the Hilbert space $L_2(I)$.

Problem 6.   Let P and Q be two linear operators in a normed space. Let a be a nonzero real or complex number and let I be the identity operator. Show that if all iterated operators Q^m exist then the relation

$$PQ - QP = aI \tag{1}$$

cannot be satisfied by two bounded operators P, Q in a normed space.

Solution.   We assume that the iterated operators $Q^m (m = 0, 1, 2 \ldots$ are meaningful. From (1) it follows that P and Q can neither be the null operator nor a constant. Consequently,

$$\|P\| \neq 0, \quad \|Q\| \neq 0$$

where $\|\cdot\|$ denotes the norm. From (1) we find by induction that

$$PQ^n - Q^nP = anQ^{n-1}. \tag{2}$$

For $n = 1$, (2) is obviously true. It follows then that

$$PQ^{n+1} - Q^{n+1}P = (PQ^n - Q^nP)Q + Q^n(PQ - QP)$$
$$= anQ^{n-1}Q + Q^naI = a(n+1)Q^n \tag{3}$$

which completes the induction. We assume now that P and Q are bounded operators with the norms $\|P\|$ and $\|Q\|$. Taking norms on both sides of (2) for $n = 1, 2, \ldots$ we find

$$|a|n\|Q^{n-1}\| = \|PQ^n - Q^nP\| \leq \|P\|\|Q^n\| + \|Q^n\|\|P\| = 2\|P\|\|Q^n\|.$$

Thus

$$|a|n\|Q^{n-1}\| \leq 2\|P\|\|Q^{n-1}\|\|Q\|.$$

Let N be an integer with

$$N > \frac{2}{|a|}\|P\|\|Q\|.$$

We have to study two cases:

Case 1: Let

$$\|Q^{N-1}\| \neq 0.$$

This result is in contradiction with (3) for $n = N$.

Case 2: Let

$$\|Q^{N-1}\| = 0.$$

Owing to

$$|a|n\|Q^{n-1}\| \leq 2\|P\|\|Q^n\|$$

for all $n = 1, 2, 3, \ldots$ we obtain

$$\|Q^{N-2}\| = 0, \quad \|Q^{N-3}\| = 0, \ldots \qquad \|Q^1\| = 0.$$

Again we find a contradiction.

Remark 1. *Owing to this result there are no two finite dimensional matrices P and Q which satisfy (1) with $a \neq 0$. This also follows from the fact that*

$$\operatorname{tr}([A, B]) = 0$$

for $n \times n$ matrices A and B over the field of the complex numbers. Here $[,]$ denotes the commutator.

Remark 2.   *There are unbounded operators which satisfy (1). Consider the infinite-dimensional matrix*

$$b := \begin{pmatrix} 0 & \sqrt{1} & 0 & 0 & \cdots \\ 0 & 0 & \sqrt{2} & 0 & \cdots \\ 0 & 0 & 0 & \sqrt{3} & \cdots \\ & & \ddots & & \ddots \\ & & & \ddots & & \ddots \\ & & & & \ddots & & \ddots \end{pmatrix}.$$

Let b^T be the transpose of b. Then

$$[b, b^T] = I$$

where I is the infinite-dimensional unit matrix and $[,]$ denotes the commutator. The operator $b^T b$ is given by the infinite-dimensional diagonal matrix

$$b^T b = \mathrm{diag}(0, 1, 2, \ldots).$$

Problem 7.   Let F be a closed set on a complete metric space X. A *contracting mapping* is a mapping $f : F \to F$ such that

$$d(f(x), f(y)) \leq k d(x, y), \quad 0 \leq k < 1, \quad d \text{ distance in} X.$$

One also says that f is Lipschitzian of order $k < 1$.

(i) Prove the following theorem: A contracting mapping f has strictly one fixed point, i.e. there is one and only one point x^* such that $x^* = f(x^*)$.

(ii) Apply the theorem to the linear equation $A\mathbf{x} = \mathbf{b}$, where A is an $n \times n$ matrix over the real numbers and $\mathbf{x} = (x_1, x_2, \ldots, x_n)^T$.

Solution.   (i) The proof is by iteration. Let $x_0 \in F$. Then

$$f(x_0) \in F, \ldots, f^{(n)}(x_0) = f(f^{(n-1)}(x_0)) \in F$$

and

$$d(f^{(n)}(x_0), f^{(n-1)}(x_0)) \leq kd(f^{(n-1)}(x_0), f^{(n-2)}(x_0))$$
$$\leq \cdots \leq k^{n-1}d(f(x_0), x_0).$$

Since $k < 1$ the sequence $f^{(n)}$ is a Cauchy sequence and tends to a limit x^* when n tends to infinity

$$x^* = \lim_{n \to \infty} f^{(n)}(x_0) = \lim_{n \to \infty} f(f^{(n-1)}(x_0)) = f(x^*).$$

The uniqueness of x^* results from the defining property of contracting mappings: Assume that there is another point y^* such that $y^* = f(y^*)$, then on the one hand

$$d(f(y^*), f(x^*)) = d(y^*, x^*)$$

but on the other hand $d(f(y^*), f(x^*)) \leq kd(y^*, x^*)$, $k < 1$. Therefore $d(y^*, x^*) = 0$ and $y^* = x^*$.

ii) We set $c_{jk} := -a_{jk} + \delta_{jk}$, where a_{jk} are the matrix elements of A and δ_{jk} is the Kronecker delta. Then the equation $A\mathbf{x} = \mathbf{b}$ takes the form $\mathbf{x} = C\mathbf{x} + \mathbf{b}$. If

$$\sum_{k=1}^{n} |c_{jk}| < 1, \quad j = 1, 2, \ldots, n$$

then the theorem can be applied.

Problem 8. Let $f_1, f_2, \ldots, f_n$ be continuous real-valued functions on the interval $[a, b]$. Show that the set

$$\{f_1, f_2, \ldots, f_n\}$$

linearly dependent on $[a, b]$ if and only if

$$\det \left(\int_a^b f_i(x) f_j(x) dx \right) = 0.$$

Solution. Let A be the matrix with the entries

$$A_{ij} := \int_a^b f_i(x) f_j(x) dx.$$

If the determinant of the matrix A vanishes, the matrix A is singular. Let $\mathbf{a}$ be a nonzero (column) n-vector with $A\mathbf{a} = \mathbf{0}$. Then

$$0 = \mathbf{a}^T A \mathbf{a} = \sum_{i=1}^{n} \sum_{j=1}^{n} \int_{a}^{b} a_i f_i(x) a_j f_j(x) dx = \int_{a}^{b} \left(\sum_{i=1}^{n} a_i f_i(x) \right)^2 dx .$$

Since the f_i's are continuous functions, the linear combinations

$$\sum_{i=1}^{n} a_i f_i$$

must vanish identically. Hence, the set $\{f_i\}$ is linearly dependent on $[a, b]$. Conversely, if $\{f_i\}$ is linearly dependent, some f_i can be expressed as a linear combination of the rest, so some row of A is a linear combination of the rest and A is singular.

Problem 9.   Consider the Hilbert space $L_2[0, \pi]$ and the function $f(x) = \sin(x)$. Find a and b $(a, b \in \mathbf{R})$ such that

$$\| f(x) - (ax^2 + bx) \|$$

is a minimum. The norm in the Hilbert space $L_2[0, \pi]$ is induced by the scalar product. Hence

$$\| f(x) - (ax^2 + bx) \|^2 = \int_{0}^{\pi} (f(x) - (ax^2 + bx))^2 dx .$$

Solution.   We have

$$h(a, b) := \int_{0}^{\pi} (\sin(x) - (ax^2 + bx))^2 dx .$$

From the conditions

$$\frac{\partial h}{\partial a} = 0, \quad \frac{\partial h}{\partial b} = 0,$$

we obtain

$$\int_{0}^{\pi} x^2 (\sin(x) - (ax^2 + bx)) dx = 0, \tag{1a}$$

$$\int_{0}^{\pi} x (\sin(x) - (ax^2 + bx)) dx = 0. \tag{1b}$$

From Eqs. (1a) and (1b) we obtain

$$a \int_0^\pi x^4 dx + b \int_0^\pi x^3 dx = \int_0^\pi x^2 \sin(x) dx \,,$$

$$a \int_0^\pi x^3 dx + b \int_0^\pi x^2 dx = \int_0^\pi x \sin(x) dx \,.$$

Integrating yields

$$\frac{\pi^5 a}{5} + \frac{\pi^4 b}{4} = \pi^2 - 4 \,,$$

$$\frac{\pi^4 a}{4} + \frac{\pi^3 b}{3} = \pi \,.$$

This is a system of linear equations for a and b. Solving for a and b we find

$$a = \frac{20}{\pi^3} - \frac{320}{\pi^5} \,, \quad b = \frac{240}{\pi^4} - \frac{12}{\pi^2} \,.$$

Problem 10. Consider the Hilbert space $L_2(\mathbf{R})$. Let $\phi \in L_2(\mathbf{R})$ be a real-valued function with

$$\int_{\mathbf{R}} \phi(x) dx = 1 \,,$$

$$\int_{\mathbf{R}} \phi(x) \phi(x - n) dx = \delta_{n0}, n \in \mathbf{Z} \,.$$

Assume that

$$\phi(x) = \sum_{k=0}^{M-1} c_k \phi(2x - k)$$

where $c_k \in \mathbf{R}$.

i) Show that

$$\sum_{k=0}^{M-1} c_k = 2 \,. \tag{1}$$

ii) Show that

$$\sum_{k=0}^{M-1} c_k^2 = 2 \,. \tag{2}$$

iii) Give a function that satisfies these conditions.

Solution.   (i) We have

$$1 = \int_{\mathbf{R}} \phi(x)dx = \int_{\mathbf{R}} \sum_{k=0}^{M-1} c_k\phi(2x-k)dx = \sum_{k=0}^{M-1} c_k \int_{\mathbf{R}} \phi(2x-k)dx\,.$$

Using the transformation

$$y = 2x - k$$

and

$$dy = 2dx\,,$$

we find

$$1 = \frac{1}{2} \sum_{k=0}^{M-1} c_k \int_{\mathbf{R}} \phi(y)dy = \frac{1}{2} \sum_{k=0}^{M-1} c_k\,.$$

Thus Eq. (1) follows.

(ii) To prove Eq. (2) we start from

$$1 = \int_{\mathbf{R}} \phi(x)\phi(x)dx = \int_{\mathbf{R}} \left(\sum_{k=0}^{M-1} c_k\phi(2x-k) \right) \left(\sum_{j=0}^{M-1} c_j\phi(2x-j) \right) dx\,.$$

Thus

$$1 = \sum_{k=0}^{M-1} \sum_{j=0}^{M-1} c_k c_j \int_{\mathbf{R}} \phi(2x-k)\phi(2x-j)dx\,.$$

Using $y = 2x - k$ and $dy = 2dx$, we obtain

$$1 = \frac{1}{2} \sum_{k=0}^{M-1} \sum_{j=0}^{M-1} c_k c_j \int_{\mathbf{R}} \phi(y)\phi(y+k-j)dy = \frac{1}{2} \sum_{k=0}^{M-1} c_k^2\,.$$

(iii) The function

$$\phi(x) = \begin{cases} 1 & \text{for } 0 \le x \le 1, \\ 0 & \text{otherwise} \end{cases}$$

satisfies the conditions given above.

Remark.   *The function ϕ plays an important role in wavelet theory. The function ϕ is called the* scaling function.

What is desired is a function ψ which is also orthogonal to its dilations, or scales, i.e.,

$$\int_{\mathbf{R}} \psi(x)\psi(2x - k)dx = 0\,.$$

Such a function ψ (the so-called associated *wavelet function*) does exist and is given by

$$\psi(x) = \sum_{k=1}^{M}(-1)^k c_{1-k}\phi(2x - k)$$

which is dependent on the solution of ϕ. Periodic boundary conditions are used

$$c_k \equiv c_{k+nM}\,.$$

Chapter 19

Combinatorics

Problem 1. (i) In how many ways f_n can a group of n persons arrange themselves in a row of n chairs?

(ii) In how many ways f_n can a group of n persons arrange themselves around a circular table.

Solution. (i) The n persons can arrange themselves in a row in

$$n(n-1)(n-2)\cdots 2 \cdot 1 = n! \tag{1}$$

ways. In other words: one person can sit in n different chairs. Therefore

$$f_n = n f_{n-1} \tag{2}$$

with $n = 2, 3, \ldots$ and

$$f_1 = 1. \tag{3}$$

The solution of the difference equation (2) with the initial condition (3) is given by Eq. (1).

(ii) One person can sit in any place at the circular table. The other $n -$ persons can then arrange themselves in

$$f_n = (n-1)(n-2)\cdots 2 \cdot 1 = (n-1)!$$

In other words n objects can be arranged in a circle in $(n-1)!$ ways.

Problem 2.   What is the number A_n of ways of going up n steps, if we may take one or two steps at a time? Determine

$$\sum_{n=0}^{\infty} A_n x^n .$$

Solution.   We may begin by taking one or two steps. In the first case, we have A_{n-1} possibilities to continue; in the second, we have A_{n-2} possibilities. Thus

$$A_n = A_{n-1} + A_{n-2} \tag{1}$$

where

$$A_1 = 1, \quad A_2 = 2 .$$

The sequence A_n is the sequence of the Fibonacci numbers. We set $A_0 = 1$ and

$$f(x) = \sum_{n=0}^{\infty} A_n x^n$$

as the generating function. Then

$$x f(x) = \sum_{n=1}^{\infty} A_{n-1} x^n ,$$

$$x^2 f(x) = \sum_{n=2}^{\infty} A_{n-2} x^n ,$$

and using (1) we find

$$f(x) - x f(x) - x^2 f(x) = A_0 + (A_1 - A_0)x + \sum_{n=2}^{\infty}(A_n - A_{n-1} - A_{n-2})x^n .$$

Thus

$$f(x) - x f(x) - x^2 f(x) = A_0 = 1 .$$

Consequently, the generating function is given by

$$f(x) = \frac{1}{1 - x - x^2} .$$

To obtain an explicit formula for A_n we write this as

$$f(x) = \frac{1}{\sqrt{5}x} \left(\frac{1}{1 - \frac{1+\sqrt{5}}{2}x} - \frac{1}{1 - \frac{1-\sqrt{5}}{2}x} \right)$$

$$= \frac{1}{\sqrt{5}x} \sum_{k=0}^{\infty} \left(\left(\frac{1+\sqrt{5}}{2}x \right)^k - \left(\frac{1-\sqrt{5}}{2}x \right)^k \right)$$

$$= \frac{1}{\sqrt{5}} \sum_{n=0}^{\infty} \left(\left(\frac{1+\sqrt{5}}{2} \right)^{n+1} - \left(\frac{1-\sqrt{5}}{2} \right)^{n+1} \right) x^n .$$

Therefore

$$A_n = \frac{1}{\sqrt{5}} \left(\left(\frac{1+\sqrt{5}}{2} \right)^{n+1} - \left(\frac{1-\sqrt{5}}{2} \right)^{n+1} \right)$$

where $n = 0, 1, 2, \ldots$.

Problem 3.   We have n Rand. Every day we buy exactly one of the following products: pretzel (1 Rand), candy (2 Rand), ice-cream (2 Rand). What is the number B_n of possible ways of spending all the money?

Solution.   On the first day, we have three choices: we may buy a pretzel, in which case we have B_{n-1} further possible ways to spend the remaining $n - 1$ Rand; or we may buy candy for 2 Rand, and then we can spend the rest in B_{n-2} ways; similarly we have B_{n-2} possibilities if we first buy an ice-cream. Thus we find the linear difference equation with constant coefficients

$$B_n = B_{n-1} + 2B_{n-2}(n \geq 3) . \tag{1}$$

The difference equation (1) is of second order. Thus we need two initial conditions. Obviously,

$$B_1 = 1, \quad B_2 = 3 \tag{2}$$

are the initial values of the linear difference equation (1). The first few terms are

$$B_3 = 5, \quad B_4 = 11, \quad B_5 = 21, \quad B_6 = 43 .$$

Since (1) is a linear difference equation with constant coefficients we can solve it with the ansatz

$$B_n = ar^n.$$

We obtain the algebraic equation

$$r^2 = r + 2.$$

Consequently, the solution of (1) with the initial conditions (2) is given by

$$B_n = \frac{1}{3}(2^{n+1} + (-1)^n).$$

Problem 4. Let $A_1, A_2, \ldots A_p$ be finite sets. Let $|A_k|$ be the number of elements of A_k. Show that

$$\left| \bigcup_{j=1}^{p} A_j \right| = \sum_{j=1}^{p} |A_j| - \sum_{1 \le j < k \le p} |A_j \cap A_k| + \cdots + (-1)^{p+1} \left| \bigcap_{j=1}^{p} A_j \right|. \quad (1)$$

Solution. The proof applies induction on $p \ge 2$. For $p = 2$ we find that (1) takes the form

$$|A_1 \cup A_2| = |A_1| + |A_2| - |A_1 \cap A_2|,$$

which is obviously true. Suppose the equation is true for each union of at most $p - 1$ sets. It follows that

$$|A_1 \cup A_2 \cdots \cup A_p| = |A_1 \cup A_2 \cdots \cup A_{p-1}| + |A_p| - |(A_1 \cup A_2 \cdots \cup A_{p-1}) \cap A_p|.$$

Applying the *distributive law for intersections* of sets we find

$$(A_1 \cup A_2 \cdots \cup A_{p-1}) \cap A_p = (A_1 \cap A_p) \cup (A_2 \cap A_p) \cup \cdots \cup (A_{p-1} \cap A_p)$$

and from the inductive hypothesis it follows that

$$|A_1 \cup A_2 \cdots \cup A_p| = \sum_{1 \le j < p} |A_j| - \sum_{1 \le j < k < p} |A_j \cap A_k| + \cdots (-1)^p \left| \bigcap_{j=1}^{p-1} A_j \right|$$

$$+ |A_p| - \sum_{1 \le j < p} |A_j \cap A_p| + \cdots (-1)^{p+1} \left| \bigcap_{j=1}^{p} A_j \right|$$

where we have used the *idempotent law of intersection* in the form

$$(A_j \cap A_p) \cap (A_k \cap A_p) = A_j \cap A_k \cap A_p, \quad \cdots \bigcap_{j=1}^{p-1} (A_j \cap A_p) = \bigcap_{j=1}^{p} A_j .$$

By regrouping terms, we obtain (1).

Remark. *Equation (1) is called the* principle of inclusion and exclusion.

Problem 5. Show that the number of arrangements N of a set of n objects in p boxes such that the jth box contains n_j objects, for $j = 1, \ldots, p$ is equal to

$$N = \frac{n!}{n_1! n_2! \cdots n_p!}$$

where $n_j \geq 0$ and

$$n_1 + n_2 + \cdots + n_p = n .$$

Solution. The n_1 objects in the first box can be chosen in

$$\binom{n}{n_1}$$

ways, the n_2 objects in the second box can be chosen from the $n - n_1$ remaining objects in

$$\binom{n - n_1}{n_2}$$

ways, etc.. The total number of arrangements is equal to

$$\binom{n}{n_1} \binom{n - n_1}{n_2} \binom{n - n_1 - n_2}{n_3} \cdots \binom{n - n_1 - \cdots - n_{p-1}}{n_p}$$

$$\equiv \frac{n!}{n_1! n_2! \cdots n_p!} .$$

Chapter 20

Convex Sets and Functions

Problem 1. A subset C of $\mathbf{R}^n$ is said to be *convex* if for any $\mathbf{a}$ and $\mathbf{b}$ in C and any θ in $\mathbf{R}$, $0 \le \theta \le 1$, the n-tuple $\theta\mathbf{a} + (1-\theta)\mathbf{b}$ also belongs to C. In other words, if $\mathbf{a}$ and $\mathbf{b}$ are in C, then

$$\{\theta\mathbf{a} + (1-\theta)\mathbf{b} : 0 \le \theta \le 1\} \subset C.$$

Let B^n be the unit ball in $\mathbf{R}^n$, i.e.

$$B^n := \left\{ (a_1, a_2, \ldots, a_n) \in \mathbf{R}^n : \sum_{j=1}^{n} a_j^2 \le 1 \right\}.$$

Show that B^n is convex.

Solution. Let

$$\mathbf{a} \in B^n, \quad \mathbf{b} \in B^n.$$

Let

$$c_j := \theta a_j + (1-\theta)b_j$$

where $j = 1, 2, \ldots, n$. Then we have

$$\sum_{j=1}^{n} c_j^2 = \sum_{j=1}^{n} (\theta a_j + (1-\theta)b_j)^2 = \theta^2 \sum_{j=1}^{n} a_j^2 + (1-\theta)^2 \sum_{j=1}^{n} b_j^2 + 2\theta(1-\theta) \sum_{j=1}^{n} a_j b_j.$$

Thus

$$\sum_{j=1}^{n} c_j^2 \le \theta^2 + (1-\theta)^2 + 2\theta(1-\theta) \sum_{j=1}^{n} a_j b_j \,.$$

Next we prove that

$$\sum_{j=1}^{n} a_j b_j \le 1 \,.$$

Since

$$0 \le \sum_{j=1}^{n} (a_j - b_j)^2 = \sum_{j=1}^{n} a_j^2 + \sum_{j=1}^{n} b_j^2 - 2 \sum_{j=1}^{n} a_j b_j \,,$$

we have

$$0 \le 2 - 2 \sum_{j=1}^{n} a_j b_j, \quad \text{or } 1 - \sum_{j=1}^{n} a_j b_j \ge 0 \,.$$

It follows that

$$\sum_{j=1}^{n} c_j^2 \le \theta^2 + (1-\theta)^2 + 2\theta(1-\theta) = (\theta + (1-\theta))^2 = 1 \,.$$

Problem 2.   Let C be a convex subset of $\mathbf{R}^n$. Let f be a real-valued function with domain C. If for each $\mathbf{x}$ and $\mathbf{y}$ in C and each θ, $0 \le \theta \le 1$, the inequality

$$f(\theta \mathbf{x} + (1-\theta)\mathbf{y}) \le \theta f(\mathbf{x}) + (1-\theta)f(\mathbf{y})$$

holds then f is said to be a *convex function*. If $-f$ is convex then f is called a *concave function*.

Let $b > a$ and consider the function

$$f(x) = x^2 \tag{1}$$

in this interval. Show that f is convex.

Solution.   We have to evaluate the difference

$$d := \theta f(x) + (1-\theta)f(y) - f(\theta x + (1-\theta)y) \,.$$

We have to show that $d \geq 0$ for any $\mathbf{x}$ and $\mathbf{y}$ in C and any θ with $0 \leq \theta \leq 1$. From (1) we find

$$
\begin{aligned}
d &= \theta x^2 + (1 - \theta)y^2 - (\theta x + (1 - \theta)y)^2 \\
&= \theta x^2 + (1 - \theta)y^2 - \theta^2 x^2 - (1 - \theta)^2 y^2 - 2\theta(1 - \theta)xy \\
&= x^2(\theta - \theta^2) + y^2((1 - \theta) - (1 - \theta)^2) - 2\theta(1 - \theta)xy \\
&= x^2\theta(1 - \theta) + y^2\theta(1 - \theta) - 2\theta(1 - \theta)xy \\
&= \theta(1 - \theta)(x^2 + y^2 - 2xy).
\end{aligned}
$$

Thus

$$ d = \theta(1 - \theta)(x - y)^2. $$

Since

$$ \theta \geq 0, \quad 1 - \theta \geq 0 $$

and

$$ (x - y)^2 \geq 0, $$

it follows that

$$ d \geq 0. $$

Problem 3. Let E^n be the n-dimensional Euclidean space. Let S be a closed convex set in E_n and $\mathbf{y} \notin S$.

i) Show that there exists a unique point $\bar{\mathbf{x}} \in S$ with a minimum distance from $\mathbf{y}$.

ii) Show that $\bar{\mathbf{x}}$ is the minimizing point if and only if

$$ (\mathbf{x} - \bar{\mathbf{x}})^T(\bar{\mathbf{x}} - \mathbf{y}) \geq 0 $$

for all $\mathbf{x} \in S$.

Solution. (i) Let

$$ \inf\{\|\mathbf{y} - \mathbf{x}\| : \mathbf{x} \in S\} = \gamma > 0. $$

There exists a sequence $\{\mathbf{x}_k\}$ in S such that $\|\mathbf{y} - \mathbf{x}_k\| \to \gamma$. We show that $\{\mathbf{x}_k\}$ has a limit $\bar{\mathbf{x}} \in S$ by showing that $\{\mathbf{x}_k\}$ is a *Cauchy sequence*. By the parallelogram law, we have

$$ \|\mathbf{x}_k - \mathbf{x}_m\|^2 = 2\|\mathbf{x}_k - \mathbf{y}\|^2 + 2\|\mathbf{x}_m - \mathbf{y}\|^2 - \|\mathbf{x}_k + \mathbf{x}_m - 2\mathbf{y}\|^2. $$

Therefore

$$\|x_k - x_m\|^2 = 2\|x_k - y\|^2 + 2\|x_m - y\|^2 - 4\left\|\frac{x_k + x_m}{2} - y\right\|^2.$$

Note that $(x_k + x_m)/2 \in S$, and by definition of γ we have

$$\left\|\frac{x_k + x_m}{2} - y\right\|^2 \geq \gamma^2.$$

Therefore

$$\|x_k - x_m\|^2 \leq 2\|x_k - y\|^2 + 2\|x_m - y\|^2 - 4\gamma^2.$$

By choosing k and m sufficiently large, $\|x_k - y\|^2$ and $\|x_m - y\|^2$ can be made sufficiently close to γ^2, hence $\|x_k - x_m\|^2$ can be made sufficiently close to zero. Therefore $\{x_k\}$ is a Cauchy sequence and has a limit $\bar{x}$. Since S is closed, $\bar{x} \in S$. To show uniqueness, suppose that there is an $\bar{x}' \in S$ such that

$$\|y - \bar{x}\| = \|y - \bar{x}'\| = \gamma.$$

As a result of the convexity of S,

$$\frac{1}{2}(\bar{x} + \bar{x}') \in S.$$

By the Schwarz inequality, we obtain

$$\left\|y - \frac{\bar{x} + \bar{x}'}{2}\right\| \leq \frac{1}{2}\|y - \bar{x}\| + \frac{1}{2}\|y - \bar{x}'\| = \gamma.$$

If strict inequality holds, we violate the definition of γ. Therefore equality holds, and we must have

$$y - \bar{x} = \lambda(y - \bar{x}')$$

for some λ. Since

$$\|y - \bar{x}\| = \|y - \bar{x}'\| = \gamma$$

it follows that $|\lambda| = 1$. Clearly, $\lambda \neq -1$, because otherwise $y = (\bar{x} + \bar{x}')/2 \in S$, contradicting the assumption that $y \notin S$. So $\lambda = 1$, $\bar{x}' = \bar{x}$, and uniqueness is established.

(ii) We need to show that

$$(x - \bar{x})^T(\bar{x} - y) \geq 0$$

for all $\mathbf{x} \in S$ is both a necessary and sufficient condition for $\bar{\mathbf{x}}$ to be the point in S closest to $\mathbf{y}$. To prove sufficiency, let $\mathbf{x} \in S$. Then,

$$\|\mathbf{y} - \mathbf{x}\|^2 = \|\mathbf{y} - \bar{\mathbf{x}} + \bar{\mathbf{x}} - \mathbf{x}\|^2 = \|\mathbf{y} - \bar{\mathbf{x}}\|^2 + \|\bar{\mathbf{x}} - \mathbf{x}\|^2 + 2(\bar{\mathbf{x}} - \mathbf{x})^T(\mathbf{y} - \bar{\mathbf{x}}).$$

Since $\|\bar{\mathbf{x}} - \mathbf{x}\|^2 \geq 0$ and $(\bar{\mathbf{x}} - \mathbf{x})^T(\mathbf{y} - \bar{\mathbf{x}}) \geq 0$ by assumption, $\|\mathbf{y} - \mathbf{x}\|^2 \geq \|\mathbf{y} - \bar{\mathbf{x}}\|^2$, and $\bar{\mathbf{x}}$ is the minimizing point. Conversely, assume that $\|\mathbf{y} - \mathbf{x}\|^2 \geq \|\mathbf{y} - \bar{\mathbf{x}}\|^2$ for all $\mathbf{x} \in S$. Let $\mathbf{x} \in S$ and note that $\bar{\mathbf{x}} + \lambda(\mathbf{x} - \bar{\mathbf{x}}) \in S$ for $\lambda > 0$ and sufficiently small. Therefore,

$$\|\mathbf{y} - \bar{\mathbf{x}} - \lambda(\mathbf{x} - \bar{\mathbf{x}})\|^2 \geq \|\mathbf{y} - \bar{\mathbf{x}}\|^2. \tag{1}$$

Also

$$\|\mathbf{y} - \bar{\mathbf{x}} - \lambda(\mathbf{x} - \bar{\mathbf{x}})\|^2 = \|\mathbf{y} - \bar{\mathbf{x}}\|^2 + \lambda^2 \|\mathbf{x} - \bar{\mathbf{x}}\|^2 + 2\lambda(\mathbf{x} - \bar{\mathbf{x}})^T(\bar{\mathbf{x}} - \mathbf{y}). \tag{2}$$

From (1) and (2) we obtain

$$\lambda^2 \|\mathbf{x} - \bar{\mathbf{x}}\|^2 + 2\lambda(\mathbf{x} - \bar{\mathbf{x}})^T(\bar{\mathbf{x}} - \mathbf{y}) \geq 0$$

for all $\lambda > 0$ and sufficiently small. Dividing by $\lambda > 0$ and letting $\lambda \to 0$, the result follows.

Problem 4. (i) Show that the intersection of convex sets is convex.

(ii) Show that the Cartesian product $I_1 \times \cdots \times I_n \subset \mathbf{R}^n$ of the intervals

$$I_j := \{x \in \mathbf{R} : a_j \leq x \leq b_j\}, \quad j = 1, 2, \ldots, n$$

is convex.

Solution. (i) Let S and T be two convex sets. Let a and b be in $S \cap T$. Then for any $0 \leq \theta \leq 1$, $\theta a + (1 - \theta)b$ is in S, since S is convex, and $\theta a + (1 - \theta)b$ is in T, since T is convex. Hence $(\theta a + (1 - \theta)b) \in S \cap T$. Therefore $S \cap T$ is convex.

(ii) Let

$$\mathbf{x} = (x_1, \ldots, x_n), \quad \mathbf{y} = (y_1, \ldots, y_n)$$

with

$$a_j \leq x_j \leq b_j, \quad a_j \leq y_j \leq b_j, \quad j = 1, \ldots, n.$$

Let $0 \leq \theta \leq 1$. Then if

$$\mathbf{z} := \theta \mathbf{x} + (1 - \theta)\mathbf{y},$$

we have

$$z_j = \theta x_j + (1 - \theta)y_j \le \theta b_j + (1 - \theta)b_j = b_j, \quad j = 1,\dots,n,$$
$$z_j = \theta x_j + (1 - \theta)y_j \ge \theta a_j + (1 - \theta)a_j = a_j, \quad j = 1,\dots,n.$$

Hence

$$a_j \le z_j \le b_j, \quad j = 1,\dots,n$$

and

$$\mathbf{z} \in I_1 \times \cdots \times I_n.$$

Thus $I_1 \times \cdots \times I_n$ is convex.

Chapter 21

Optimization

Problem 1. A number of optimization problems may be cast in the form

$$\text{Min } E = c_1 x_1 + c_2 x_2 + \cdots + c_n x_n \equiv \sum_{j=1}^{n} c_j x_j$$

with linear constraints

$$\sum_{j=1}^{n} a_{ij} x_j (\leq, =, \geq) b_i, \quad i = 1, 2, \ldots, m$$

$$x_j \geq 0, \quad j = 1, 2, \ldots, n$$

where the coefficients c_j and a_{ij} are constants and the notation $(\leq, =, \geq)$ signifies that any of the three possibilities may hold in any constraint. This is the standard linear programming problem.

Find the minimum of

$$E = -5x_1 - 4x_2 - 6x_3$$

subject to

$$x_1 + x_2 + x_3 \leq 100,$$

$$3x_1 + 2x_2 + 4x_3 \leq 210,$$

$$3x_1 + 2x_2 \leq 150,$$

$$x_1, x_2, x_3 \geq 0.$$

Solution.   We convert to equalities by introducing three nonnegative *slack variables* x_4, x_5, x_6.

$$x_1 + x_2 + x_3 + x_4 = 100$$

$$3x_1 + 2x_2 + 4x_3 + x_5 = 210 \tag{1}$$

$$3x_1 + 2x_2 + x_6 = 150$$

$$5x_1 + 4x_2 + 6x_3 + E = 0 \tag{2}$$

where

$$x_1, x_2, x_3, x_4, x_5, x_6 \geq 0.$$

A basic feasible solution to this system is

$$x_4 = 100, \quad x_5 = 210, \quad x_6 = 150, \quad x_1 = 0, \quad x_2 = 0, \quad x_3 = 0.$$

The nonzero variables x_4, x_5, x_6 are the basic variables at this stage. Thus $E = 0$. Now, computing the gradient of E from (2),

$$\frac{\partial E}{\partial x_1} = -5, \quad \frac{\partial E}{\partial x_2} = -4, \quad \frac{\partial E}{\partial x_3} = -6$$

so that improvement in E can be obtained by increasing x_1, x_2, and/or x_3. Since the magnitude of the gradient in the x_3 direction is greatest we move in x_3-direction. We retain x_1 and x_2 as nonbasic (zero). From (1) we obtain

$$x_3 + x_4 = 100,$$

$$4x_3 + x_5 = 210.$$

Either x_4 or x_5 must go to zero (since x_3 will be nonzero) while the other remains nonnegative. If x_4 goes to zero, $x_3 = 100$ and $x_5 = -190$, while if x_5 goes to zero $x_3 = 52.5$ and $x_4 = 47.5$. That is

$$x_3 = \min\left(\frac{100}{1}, \frac{210}{4}\right) = 52.5$$

and x_5 is to be eliminated (set to zero) as x_3 is introduced. The calculation is one of dividing the right-hand column by the coefficient of x_3 and comparing. The new basic variables, x_4, x_3, and x_6 each appear in only a single equation including (2). This is done by dividing the second equation by the coefficient of x_3, then multiplying it by the coefficient of x_3 in each of the other equations, and subtracting to obtain an equivalent set of equations

The second equation is chosen because it is the one in which x_5 appears. Thus we find

$$\frac{1}{4}x_1 + \frac{1}{2}x_2 + x_4 - \frac{1}{4}x_5 = 47\frac{1}{2}$$

$$\frac{3}{4}x_1 + \frac{1}{2}x_2 + x_3 + \frac{1}{4}x_5 = 52\frac{1}{2}\,, \tag{3}$$

$$3x_1 + 2x_2 + x_6 = 150\,,$$

and

$$\frac{1}{2}x_1 + x_2 - \frac{3}{2}x_5 + E = -315\,. \tag{4}$$

The basic feasible solution is

$$x_1 = 0\,, \quad x_2 = 0\,, \quad x_5 = 0\,, \quad x_4 = 47\frac{1}{2}\,, \quad x_3 = 52\frac{1}{2}\,, \quad x_6 = 150\,.$$

From (4)

$$E = -315 - \frac{1}{2}x_1 - x_2 + \frac{3}{2}x_5$$

so that the value of E in the basic variables is -315. Repeating the above procedure, we now find that the largest positive coefficient in (4) is that of x_2, and so x_2 is to be the new basic (nonzero) variable. The variable to be eliminated is found from the coefficients of the basic variable in system (3) in the form

$$x_2 = \min\left(\frac{47\frac{1}{2}}{\frac{1}{2}}, \frac{52\frac{1}{2}}{\frac{1}{2}}, \frac{150}{2}\right) = 75$$

which corresponds to eliminating x_6. Thus, we now use the Gauss–Jordan procedure to obtain x_2 in the third equation only. The result is

$$-\frac{1}{2}x_1 + x_4 - \frac{1}{4}x_5 - \frac{1}{4}x_6 = 10\,,$$

$$x_3 + \frac{1}{4}x_5 - \frac{1}{4}x_6 = 15\,,$$

$$\frac{3}{2}x_1 + x_2 + \frac{1}{2}x_6 = 75\,,$$

$$-x_1 - \frac{3}{2}x_5 - \frac{1}{2}x_6 + E = -390 \tag{5}$$

The basic feasible solution is then

$$x_1 = 0, \quad x_5 = 0, \quad x_6 = 0, \quad x_4 = 10, \quad x_3 = 15, \quad x_2 = 75.$$

The corresponding value of E is

$$E = -390.$$

There are no positive coefficients in (5), and so this is the minimum.

Problem 2.   Minimize

$$\left(x_1 - \frac{3}{2}\right)^2 + (x_2 - 5)^2$$

subject to

$$-x_1 + x_2 \leq 2, \quad 2x_1 + 3x_2 \leq 11, \quad x_1 \geq 0, \quad x_2 \geq 0. \tag{1}$$

Solution.   A convex polyhedral set S is represented by the four inequalities given by (1). Let S be a nonempty convex set in $\mathbf{R}^n$.

A vector $\mathbf{x} \in S$ is called an *extreme point* of S if $\mathbf{x} = \lambda\mathbf{x}_1 + (1 - \lambda)\mathbf{x}_2$ with $\mathbf{x}_1, \mathbf{x}_2 \in S$, and $\lambda \in (0,1)$ implies $\mathbf{x} = \mathbf{x}_1 = \mathbf{x}_2$.

The extreme points of S are

$$(0,0), (0,2), (1,3), \left(\frac{11}{2}, 0\right).$$

The function

$$f(x_1, x_2) = \left(x_1 - \frac{3}{2}\right)^2 + (x_2 - 5)^2$$

is a convex function, which gives the square of the distance from the point (3/2, 5).

Let S be a nonempty convex set in $\mathbf{R}^n$ and let $f : S \to \mathbf{R}$ be convex. Then $\mathbf{y}$ is called a *subgradient* of f at $\bar{\mathbf{x}}$ if

$$f(\mathbf{x}) \geq f(\bar{\mathbf{x}}) + \mathbf{y}^T(\mathbf{x} - \bar{\mathbf{x}})$$

for all $\mathbf{x} \in S$.

Let $f : \mathbf{R}^n \to \mathbf{R}$ be a convex function, and S be a nonempty convex set in $\mathbf{R}^n$. Consider the problem to minimize f subject to $\mathbf{x} \in S$. The point $\bar{\mathbf{x}} \in S$ is an optimal solution if and only if f has a subgradient $\mathbf{y}$ at $\bar{\mathbf{x}}$ such that

$$y^T(\mathbf{x} - \bar{\mathbf{x}}) \geq 0$$

for all $\mathbf{x} \in S$.

The gradient vector of f at the point $(1, 3)$ is

$$\nabla f(1, 3) = (-1, -4)^T.$$

We see geometrically that the vector $(-1, -4)$ makes an angle of $\leq \pi/4$ with each vector of the form $(x_1 - 1, x_2 - 3)$, where $(x_1, x_2) \in S$. Thus the optimality condition given above is verified.

Now, suppose that it is claimed that $(0, 0)$ is an optimal point. Note that

$$\nabla f(0, 0) = (-3, -10)^T$$

and for each nonzero $\mathbf{x} \in S$, we have $-3x_1 - 10x_2 < 0$. Hence, the origin could not be an optimal point. Moreover, we can improve f by moving from $\mathbf{0}$ in the direction $\mathbf{x} - \mathbf{0}$ for any $\mathbf{x} \in S$. In this case, the best local direction is $-\nabla f(0, 0)$, that is the direction $(3, 10)$.

Bibliography

E. J. Barbeau, *Polynomials*, Springer-Verlag, New York (1989).

M. Berger, P. Pansu, J. P. Berry and X. Saint-Raymond, *Problems in Geometry*, Springer-Verlag, New York (1984).

F. Constantinescu and E. Magyari, *Problems in Quantum Mechanics*, Pergamon Press, Oxford (1971).

J. A. Cronin, D. F. Greenberg and V. L. Telegdi, *Graduate Problems in Physics*, Addison Wesley, Reading (1967).

P. N. de Souza and J.-N. Silva, *Berkeley Problems in Mathematics*, Springer-Verlag, New York (1998).

S. Flügge, *Practical Quantum Mechanics*, Springer-Verlag, Berlin (1974).

B. Gelbaum, *Problems in Analysis*, Springer-Verlag, New York (1982).

K. Knoop, *Problem Book in the Theory of Functions*, Volume I, Dover, New York (1952).

J. G. Krzyz, *Problems in Complex Variables Theory*, Elsevier, New York (1971).

L. C. Larson, *Problem Solving Through Problems*, Springer-Verlag, New York (1983).

J. M. Rassias, *Counter Examples in Differential Equations and Related Topics*, World Scientific, Singapore (1991).

M. R. Spiegel, *Advanced Calculus*, Schaum's Outline Series, McGraw Hill, New York (1974).

M. R. Spiegel, *Finite Differences and Difference Equations*, Schaum's Outline Series, McGraw Hill, New York (1971).

M. R. Spiegel, *Complex Variables*, Schaum's Outline Series, McGraw Hill, New York (1971).

I. Tomescu, *Problems in Combinatorics and Graph Theory*, Wiley, New York (1985).

Index